D0290726

DNA REPLICATION

METHODS IN MOLECULAR BIOLOGY

Edited by

ALLEN I. LASKIN
ESSO Research and Engineering Company
Linden, New Jersey

JEROLD A. LAST
Harvard University
Cambridge, Massachusetts

VOLUME 1: Protein Biosynthesis in Bacterial Systems, *edited by Jerold A. Last and Allen I. Laskin*

VOLUME 2: Protein Biosynthesis in Nonbacterial Systems, *edited by Jerold A. Last and Allen I. Laskin*

VOLUME 3: Methods in Cyclic Nucleotide Research, *edited by Mark Chasin*

VOLUME 4: Nucleic Acid Biosynthesis, *edited by Allen I. Laskin and Jerold A. Last*

VOLUME 5: Subcellular Particles, Structures, and Organelles, *edited by Allen I. Laskin and Jerold A. Last*

VOLUME 6: Immune Response at the Cellular Level, *edited by Theodore P. Zacharia*

VOLUME 7: DNA Replication *edited by Reed B. Wickner*

DNA REPLICATION

EDITED BY

Reed B. Wickner

National Institutes of Health
Bethesda, Maryland

MARCEL DEKKER, INC. New York

MARCEL DEKKER, INC.
270 Madison Avenue, New York, New York 10016

LIBRARY OF CONGRESS CATALOG CARD NUMBER: 74-77107
ISBN: 0-8247-6202-9

Current printing (last digit):
10 9 8 7 6 5 4 3 2 1

PRINTED IN THE UNITED STATES OF AMERICA

PREFACE

This volume is a collection of methods, most recently devised, which have been found useful by those currently studying DNA replication of *Escherichia coli* and its bacteriophages, in vivo and in vitro. Included are: (1) methods of assay and purification of various enzymes, most of which are known to be involved in DNA replication, although the enzymatic activities of some are unknown; (2) several semi-in vitro (or "in vivtro" to use Bonhoeffer's expression) systems which carry out DNA replication; and (3) methods which have been of value in the study of the structure of replicating DNA molecules. Unfortunately, limitations of space have resulted in the omission of many important methods in these areas, but it is hoped that the descriptions included in this volume will be of value to those working in this field.

REED B. WICKNER

CONTRIBUTORS TO THIS VOLUME

Numbers in parentheses indicate the pages on which the authors' contributions begin.

JACK BARRY*(257), Department of Biochemical Sciences, Princeton University, Princeton, New Jersey

IRA BERKOWER (195), Department of Developmental Biology and Cancer, Division of Biological Sciences, Albert Einstein College of Medicine, Bronx, New York

H. HOFFMAN-BERLING (151), Max-Planck-Institut für Medizinische Forschung, Abteilung Molekulare Biologie, Heidelberg, West Germany

M. C. BETLACK (87), Department of Microbiology, University of California, San Francisco, California

HERBERT W. BOYER (87), Department of Microbiology, University of California, San Francisco, California

DOUGLAS BRUTLAG (187), Department of Biochemistry, Stanford University Medical Center, Stanford, California

JONATHAN O. CARLSON (231), Department of Chemistry, University of California, Berkeley, California

---

*Present address: National Academy of Sciences, Assembly of Life Sciences, Washington, D.C.

DHRUBA K. CHATTORAJ*(33), Biophysics Laboratory, University of Wisconsin, Madison, Wisconsin

KLAUS GEIDER**(187), Department of Biochemistry, Stanford University School of Medicine, Stanford, California

HOWARD M. GOODMAN (87), Department of Biochemistry and Biophysics, University of California, San Francisco, California

P. J. GREENE (87), Department of Biochemistry and Biophysics, University of California, San Francisco, California

JERARD HURWITZ (195), Department of Developmental Biology and Cancer, Division of Biological Sciences, Albert Einstein College of Medicine, Bronx, New York

ROSS B. INMAN (33), Biophysics Laboratory and Biochemistry Department, University of Wisconsin, Madison, Wisconsin

ALBRECHT KLEIN***(163), Max-Planck-Institut für Virusforschung, Tübingen, West Germany

ARTHUR KURNBERG (187), Department of Biochemistry, Stanford University School of Medicine, Stanford, California

ROBB E. MOSES (135), Department of Biochemistry, Baylor College of Medicine, Texas Medical Center, Houston, Texas

NANCY G. NOSSAL (239), Laboratory of Biological Pharmacology, National Institute of Arthritis, Metabolism and Digestive Diseases, National Institutes of Health, Bethesda, Maryland

---

*Present address: Institute of Molecular Biology, University of Oregon, Eugene, Oregon

**Present address: Max-Planck-Institut für Medizinische Forschung, Abteilung Molekulare Biologie, Heidelberg, West Germany

***Present address: Fachgruppe Allgemeine, Biologie, Universitat, West Germany

VOLKER NÜSSLEIN (163), Max-Planck-Institut für Virusforschung, Tübingen, West Germany

REIJI OKAZAKI (1), Institute of Molecular Biology, Faculty of Science, Nagoya University, Chikusa-Ku, Nagoya, Japan

RANDY SCHEKMAN (187), Graduate Student, Department of Biochemistry, Stanford University School of Medicine, Stanford, California

HAMILTON O. SMITH (71), Department of Microbiology, The Johns Hopkins University School of Medicine, Baltimore, Maryland

M. TAKANAMI (113), Institute for Chemical Research, Kyoto University, Uji, Kyoto, Japan

H. P. VOSBERG*(151), Max-Planck-Institut für Medizinische Forschung, Abteilung Molekulare Biologie, Heidelberg, West Germany

JAMES C. WANG (231), Chemistry Department, University of California, Berkeley, California

JOEL H. WEINER (187), Department of Biochemistry, Stanford University School of Medicine, Stanford, California

REED B. WICKNER**(181), Department of Developmental Biology and Cancer, Albert Einstein College of Medicine, Bronx, New York

SUE WICKNER (195,221), Department of Developmental Biology and Cancer, Division of Biological Sciences, Albert Einstein College of Medicine, Bronx, New York

WILLIAM WICKNER (187), Stanford University Department of Biochemistry, Stanford, California

---

*Present address: California Institute of Technology, Division of Biology, Pasadena, California

**Present address: Laboratory of Biochemical Pharmacology, National Institutes of Health, Bethesda, Maryland

MICHEL WRIGHT*(195,221), Department of Developmental Biology and Cancer, Division of Biological Sciences, Albert Einstein College of Medicine, Bronx, New York

*Present address: Fellow of the Jane Coffin Childs Memorial Fund for Medical Research, Laboratoire de Pharmacologie et Toxicologie Fondamentale, C.N.R.S. Toulouse, France

CONTENTS

DNA REPLICATION

Chapter 1

SHORT-CHAIN INTERMEDIATES IN DNA REPLICATION

Reiji Okazaki

Institute of Molecular Biology
Faculty of Science
Nagoya University
Chikusa-ku, Nagoya, Japan

I. INTRODUCTION

Newly replicated DNA selectively labeled by a brief radioactive pulse is isolated, after denaturation, as small molecules of DNA ("nascent fragments") [1-3]. Upon impairment of the activity of polynucleotide ligase these nascent DNA fragments accumulate and their conversion to large DNA is inhibited [3,4].

The nascent DNA fragments are produced by a 5' → 3' synthesis [5,6]. These and some other findings support the idea that DNA is replicated by a discontinuous mechanism involving synthesis and subsequent joining of short DNA fragments. Recent studies [7-12] have revealed the possibility that an even shorter RNA chain is attached to the 5' end of the nascent fragments, suggesting that RNA synthesis is involved in the initiation of the synthesis of the fragments. This chapter deals with basic methods for detection and characterization of the nascent DNA fragments (short-chain intermediates) in *Escherichia coli*, *Bacillus subtilis*, phage T4, and phage P2.

II. GROWTH OF BACTERIA AND BACTERIOPHAGES

Any conditions that allow normal bacterial and phage growth may be used. The following are our standard conditions:

A. *Escherichia coli*

Cells are grown at 37°C with shaking to a titer of 5 x 10^8 to 10^9 cells/ml in Medium A [2], a salts and glucose medium containing 0.1 M potassium phosphate buffer (pH 7.3), 1 mM $MgSO_4$, 20 mM $(NH_4)_2SO_4$, 2 μM $Fe(NH_4)_2(SO_4)_2$, and 1% glucose, or Medium A supplemented with Casamino Acids (0.5% or 1%) (Difco). For auxotrophic strains, these media are supplemented with any necessary nutrients.

B. *Bacillus subtilis*

Cells are grown at 30°C with shaking to titers of either (a) 10^8 cells/ml in modified Spizizen medium [13] containing

1.4% K_2HPO_4, 0.6% KH_2PO_4, 0.2% $(NH_4)_2SO_4$, 0.1% sodium citrate· $2H_2O$, 0.02% $MgSO_4 \cdot 7H_2O$, and 1% glucose, supplemented with required nutrient if any; or (b) 2 x 10^8 cells/ml in Tryptose broth (Difco).

C. T4 Phage

Cells of *E. coli* (a derivative of strain B or K12) are grown as above to a titer of 10^9 cells/ml, transferred to 20°C, and 10 to 20 min later infected with T4D or its derivative at a multiplicity of 10. The infected cells were incubated for 70 to 80 min at 20°C with shaking. Cell lysis occurs about 130 min after infection under these conditions.

D. P2 Phage

Cells of *E. coli* (strain C-1a, C, or their derivative) are grown at 37°C to a titer of 6 x 10^8 cells/ml in TPG-CAA medium [14] containing 10 mM NaCl, 0.1 M KCl, 0.2 M NH_4Cl, 0.1 M Tris base (Trizma base, Sigma Chemical Co.), 7 mM KH_2PO_4, 7 mM sodium pyruvate, 1 mM $MgCl_2 \cdot 6H_2O$, 0.16 mM Na_2SO_4, 1 mM $CaCl_2$, 0.4 μM $FeCl_3 \cdot 6H_2O$, 0.2% glucose, and 1% Casamino Acids and adjusted to pH 7.4. The culture is transferred to 20°C and 20 min later infected with P2 *vir*$_1$ at a multiplicity of 10 and incubated for 100 to 120 min. Under these conditions, lysis occurs about 160 min after infection.

III. LABELING OF DNA

A. Radioactive Compounds

The following commercial products are used:

1.	[methyl-^{3}H]Thymidine	~20 Ci/mmole	New England Nuclear Corp.
2.	[methyl-^{3}H]Thymidine	~50 Ci/mmole	New England Nuclear Corp.
3.	[methyl-^{3}H]Thymine	~50 Ci/mmole	New England Nuclear Corp.
4.	[2-^{14}C]Thymidine	~60 mCi/mmole	The Radiochemical Centre, Amersham
5.	[U-^{14}C]Thymidine	~520 mCi/mmole	The Radiochemical Centre, Amersham
6.	[2-^{14}C]Thymine	~60 mCi/mmole	The Radiochemical Centre, Amersham

It is recommended that ^{3}H-labeled compounds be obtained in ethanol-water and the solutions stored at -20°C. In many preparations of [^{3}H]thymine, particularly those supplied as aqueous solutions, high-molecular-weight radioactive impurities are found. The impurities are recovered with the cells after pulse labeling. They are acid-insoluble and sediment very slowly in alkaline sucrose gradients. The commercial preparations of [^{3}H]thymine are purified by paper chromatography using n-butanol-ammonia-water (6:1:3) or by gel filtration on Sephadex G 25. They are used immediately or after less than one week's storage in ethanol-water at -20°C. The formation of impurities is very slow under these conditions.

B. Pulse Labeling

1. Temperature

Pulse labeling is often performed at a low temperature (e.g., 8°, 14°, 20°C) to facilitate the specific labeling of the most recently replicated portion of DNA. Many *E. coli* strains grow even at 8°C. The late steps of T4 phage development also occur at 8°C. In uninfected and T4-infected *E. coli*, the relative rates of DNA synthesis at 8°, 14°, 20° and 37°C are approximately 1, 4, 12, and 60. Pulse labeling of *B. subtilis* is carried out at 20°, 25° or 30°C. P2-infected cells are pulse labeled at 20°C.

2. Addition of Radioactive Precursors

a. Radioactive Thymidine to thy^+ Cells. Cultures grown in medium containing no thymine or thymidine at 37°C are transferred to a desired temperature and incubated for a certain period (from 10 to 90 min). A solution (from 0.05 to 1 ml) of radioactive thymidine is then added to the cultures, being stirred or shaken, to a final concentration of 10^{-7} to 10^{-6} M. After the desired time (from 5 sec to 10 min), the pulse is terminated by the method described below.

b. Radioactive Thymidine to thy^- Cells. Cells of a thymine-requiring strain (e.g., *E. coli* 15T$^-$, JG116, and P3478, *B. subtilis* 168, and T4 *td*8-infected *E. coli* JG116) grown in medium containing 10 to 20 μM thymidine or thymine are collected by centrifugation and resuspended in a small volume of fresh medium at 0°C containing neither thymidine nor thymine. The cell suspension is poured into a large volume of medium, being stirred at a desired temperature, so that the cell density becomes equal to that of the original culture. Radioactive thymine is added to a concentration of 10^{-7} to 10^{-6} M immediately, or after a period of thymine starvation.

c. Radioactive Thymidine to thy⁻ Cells Growing in Thymine-Containing Medium. Radioactive thymidine is added to a culture growing in medium containing 10 to 20 μM thymine at a desired temperature to a final concentration of 10^{-7} to 10^{-6} M. Under these conditions, radioactive thymidine is taken up preferentially in the presence of a large amount of thymine.

d. [^{3}H]Thymine to thy⁻ Cells Growing in Thymine-Containing Medium. [^{3}H]Thymine is added to the culture growing in thymine-containing medium (10-20 μM) at a desired temperature to a final concentration of 1 μM. The concentration of unlabeled thymine at the time of pulse labeling (and hence the specific activity of [^{3}H]thymine in the medium) can be estimated by measuring the radioactivity in the medium and cells in a culture grown under identical conditions with radioactive thymine with a low specific activity.

3. Stopping

It is important to "fix" the pulse-labeled cells in order to stop the reactions involved in DNA replication instantaneously. The following methods are used to terminate the pulse labeling.

a. Ice-KCN. The pulse-labeled culture is poured onto a mixture of crushed ice and KCN, and the cells are collected by centrifugation at 0°C. The amounts of ice and KCN are adjusted so that some ice still remains after centrifugation and the final KCN concentration is about 20 mM. Incorporation of thymidine or thymine into DNA can be stopped immediately by this procedure that was used in most of our early experiments. Our recent study [15], however, indicated that the joining of nascent fragments of P2 may not be stopped completely by this method.

b. Cold Acetone. The pulse-labeled culture is poured into a larger volume of acetone at -30°C (or -60°C), and the cells are collected by centrifugation. With this method, care must be taken to keep the temperature during centrifugation below -10°C, since an appreciable portion of nascent DNA goes to the supernatant when the sample is centrifuged at temperatures above 0°C [16].

c. Ethanol-Phenol. The pulse-labeled culture is poured into an equal volume of "ethanol-phenol mixture" (at room temperature) consisting of 75% ethanol, 21 mM sodium acetate (pH 5.3), 2 mM EDTA, and 2% phenol [6,17]. The cells are collected by centrifugation for 5 min at 7000 g and 0°C. The reactions are stopped immediately. This method appears superior to the previous two methods in that it is simple and does not require any special care and that nucleases and other enzymes may be inactivated.

4. Kinetics of Incorporation

Figures 1 and 2 show the kinetics of [^{3}H]thymidine incorporation into uninfected and T4-infected *E. coli* at 14°C obtained by using various stopping methods.

5. Specificity of Labeling

The radioactivity incorporated into cold acid-insoluble fraction after pulse labeling with radioactive thymidine or thymine ordinarily represents the incorporation of these precursors into DNA. It is recommended, however, that each system be tested to be certain that the labeled acid-insoluble material is totally resistant to alkali, RNase, amylase, and proteinase but is degraded completely by treatment with DNase or hot 5%

trichloroacetic acid (TCA). The impurities found in [^{3}H]thymine preparations (see Section III, A above) are acid insoluble but are degraded by treatment with 0.3 M NaOH at 37°C to an acid-soluble form [16].

In normally growing E. coli, virtually all the pulse-labeled DNA seems to derive from the replicating region. This conclusion is supported by experiments in which cells grown in heavy medium are transferred to light medium and then pulse labeled [16,18].

6. Terminal and Uniform Labeling of the Fragments

It is possible to label the growing end of the nascent fragments by an extremely short pulse, while the fragments are labeled uniformly by a long pulse. The nascent fragments were isolated from T4-infected cells pulse labeled at 8°C with [^{3}H]thymidine for 6 or 12 sec and with [^{14}C]thymidine for 2.5 min and were separated into complementary strands. The growing point of the fragments of both strands labeled with ^{3}H was shown to be their 3' terminus by exonucleolytic degradation of these molecules in 3' → 5' and 5' → 3' directions [5] and by selective isolation and degradation of the 5'-terminal portion of the molecules [6].

C. Labeling of the Bulk of DNA

To compare the properties of nascent and bulk DNA directly, (a) cells are labeled for a long period, e.g., for one to a few generations with [^{14}C]thymine (or [^{14}C]thymidine) prior to pulse labeling with [^{3}H]thymidine or [^{3}H]thymine, or (b) cells labeled for a long period with [^{14}C]thymidine (or [^{14}C]thymine) are mixed with cells pulse labeled with [^{3}H]thymidine (or [^{3}H]thymine);

DNA is then extracted from the mixed cells. The long labeling is done with _thy⁻_ organisms simply by growing them in a culture medium containing 10 to 20 μM [^{14}C]thymidine or [^{14}C]thymine. With _thy⁺_ strains, thymine is not incorporated and thymidine is incorporated only for a short period under ordinary conditions. In the presence of 0.2 mM deoxyadenosine, however, incorporation of thymine and thymidine continues for about one generation.

D. Chase

That the pulse-labeled DNA represents an intermediate of the synthesis of chromosomal DNA can be shown by a "chase" following the pulse labeling. The chase is achieved by adding a large amount of unlabeled thymidine (or thymine) [e.g., 10^4 times more than the radioactive thymidine (or thymine)] at a desired time after the addition of the radioactive precursor following by further incubation. The effect of the chase on the incorporation is usually not immediate, presumably because of the labeled nucleotide pool in the cells.

When radioactive thymidine is added at 10^{-7} to 10^{-6} M to a culture of _thy⁺_ strains, the incorporation rate levels off shortly (within a few minutes at 20°C). This is due to the fact that a large portion of the added thymidine is used up; however, some radioactivity remaining in the medium is in the form of thymine, which can be utilized poorly. Under these conditions, the radioactivity is chased by endogenous thymidine nucleotides.

IV. EXTRACTION OF DNA

Various methods of DNA extraction may be used, with caution, to minimize breakage of DNA by shearing and by nuclease action.

The following are some specific procedures. The description is for a quantity of 5×10^9 to 10^{10} cells.

A. Extraction of Native DNA from Uninfected or Phage-Infected *E. coli* by a Modification of the Method of Berns and Thomas [19]

Cells are suspended in 1 ml of ice-cold standard saline-citrate [(SSC); 0.15 M NaCl-0.015 M sodium citrate] containing 27% sucrose and 10 mM EDTA (pH 8.2); 50 μl of lysozyme (10 mg/ml) are added. After standing at 0°C for 10 min (or 30 min), 8.95 ml of prewarmed SSC containing 27% sucrose, 10 mM EDTA, and 100 mg of sodium dodecyl sulfate are added and the mixture is incubated at 37°C for 10 min. Then 1 ml of Pronase E (Kaken Chemical, Tokyo, 10 mg/ml in SSC-1 mM EDTA) that has been preincubated for 1 hr at 37°C is added, and the incubation is continued at 37°C for 4 hr. To the digest, 10 ml of freshly distilled phenol containing 0.1 volume of 0.1 M Tris base (Trizma base, Sigma Chemical Co.) is added and the mixture is "rolled" at 60 rpm for 30 min at room temperature. The phenol layer (upper layer) is removed by pipeting; 10 ml of phenol-Tris is added and the extraction is repeated. To the aqueous layer, 20 ml of ether is added, the preparation is rolled at 60 rpm for 10 min, and the ether layer is removed. This procedure is repeated, and the aqueous layer is dialyzed against 1 to 2 liters of 20 mM Tris-HCl buffer (pH 8.0) containing 20 mM EDTA (or other appropriate buffer) at 4°C. Dialysis is continued from 1 to 2 days with several changes of buffer. The dialyzed sample is concentrated to 1 to 2 ml by pressure dialysis in the same buffer, using a Sartorius collodion membrane, or by ethanol precipitation. For ethanol precipitation, 0.03 vol of 3 M sodium acetate (pH 5.2) and 2 vol of ethanol at -20°C are added and the mixture is kept at -20°C for 5 to 15 hr. The precipitate is collected by centrifugation for 15 min at 10,000 _g_ at -10°C, dissolved quickly in 1 to 2 ml of 20 mM Tris-HCl (pH 8.0) containing 20 mM EDTA (or other appropriate buffer), and dialyzed against the same buffer.

B. Extraction of Denatured DNA from Uninfected and Phage-Infected _E. coli_ by NaOH-EDTA-Sodium N-Lauroyl Sarcosinate

Cells are suspended in 0.5 ml of 1% sodium N-Lauroyl sarcosinate containing 40 mM EDTA, and 0.5 ml of 0.4 M NaOH is added. The mixture is incubated at 37°C for 20 min and the insoluble material is removed, after chilling, by centrifugation at 14,000 _g_ for 15 min.

C. Extraction of Denatured DNA from _E. coli_ by Hot Sodium N-Lauroyl Sarcosinate-Sodium Dodecyl Sulfate

Cells are suspended in 0.5 ml of ice-cold 1% sodium N-lauroyl sarcosinate-40 mM EDTA. To the cell suspension 4.5 ml of 0.5% sodium dodecyl sulfate-10 mM NaCl-10 mM Tris-HCl (pH 7.6)-10 mM EDTA prewarmed in a boiling water bath are added. The mixture is heated in a boiling water bath for 5 min and chilled quickly in ice water for 30 to 60 sec.

D. Extraction of Denatured DNA from _B. subtilis_ by Lysozyme, Sodium N-Lauroyl Sarcosinate, and NaOH Treatment

Labeled cells are suspended in 2.5 to 5 ml of ice-cold SSC containing 20 mM EDTA. Lysozyme is added to 2 mg/ml and the suspension is incubated at 37°C for 10 min. Sodium N-lauroyl sarcosinate and NaOH are added to a final concentration of 0.5% and 0.2 M, respectively, and the mixture is incubated for 30 min at 37°C and chilled. The insoluble material is removed by centrifugation at 14,000 _g_ for 15 min.

Recovery of labeled DNA by these procedures is generally more than 90%.

V. ANALYSIS AND FRACTIONATION OF LABELED DNA BY ALKALINE SUCROSE GRADIENT SEDIMENTATION AND OTHER METHODS

A. Alkaline Sucrose Gradient Sedimentation

Preparative zone sedimentation through alkaline sucrose gradients is extensively used for size analysis of labeled DNA. A 5 to 20% linear sucrose gradient containing 0.1 M NaOH, 0.9 M NaCl, and 1 mM EDTA is made in a nitrocellulose centrifuge tube for the Spinco SW 25.1, SW 25.3, or SW 27 rotor using the device described by McConkey [20]. A "shelf" of 80% sucrose is placed on the bottom of the centrifuge tube to prevent pelleting of the fast sedimenting material. With narrow tubes (diameter 5/8 in.) for the SW 25.3 and SW 27 rotors, 0.3 ml of a DNA sample in 0.1 M NaOH-10 mM EDTA or 0.2 M NaOH-0.5% sodium N-lauroyl sarcosinate-20 mM EDTA, which contains 10^5 to 10^6 plaque-forming units of δA phage DNA (19S in 0.1 M NaOH-0.9 M NaCl) or an appropriate amount of 14C-labeled δA DNA [21] (or any other DNA with a known sedimentation coefficient) as an internal reference, is layered on top of the gradient. With larger centrifuge tubes (diameter 1 in.) for the SW 25.1 or SW 27 rotors, a 1-ml sample is loaded. Centrifugation is carried out at 22,500 or 25,000 rpm and 4°C for 12 to 20 hr. After centrifugation, the bottoms of the centrifuge tubes are punctured and constant volumes (0.3 to 1 ml) of the samples are collected.

The acid-insoluble radioactivity of each fraction is measured by the glass filter method described by Friesen [22] or by the filter paper disk technique of Bollum [23]. In some cases, the acid-insoluble material is isolated by centrifugation. To an aliquot of each fraction, 50 μl of cold solution of sonicated salmon-sperm DNA (2 mg/ml) is added, followed by the addition of cold 10% TCA to a final concentration of 5%. The mixture is centrifuged at 10,000 g for 5 min, the precipitate is dissolved in 0.3 ml of 0.2 N NaOH, reprecipitated with 1 ml of cold 6% TCA, and centrifuged again. This procedure is often repeated once (or twice) more and the final acid-insoluble precipitate is dissolved by heating in 0.3 ml of 5% TCA at 90°C and counted. In some experiments, the first acid precipitate is dissolved in 0.3 M NaOH and incubated at 37°C for overnight before reprecipitation with TCA. The final acid precipitate is treated with 5% TCA at 90°C for 30 min, and the supernatant fluid is counted. These treatments, which remove RNA and protein, are recommended if one is working with a new system and the procedure of DNA extraction used does not involve the removal of RNA and/or protein.

When unlabeled δA DNA is used as internal reference, fractions in the region where the infectious circular δA DNA is expected are also assayed for the infectivity to speroplasts of a F+ derivative of *E. coli* W3110. The spheroplast assay is carried out as described by Sinsheimer [24].

The sedimentation pattern of the labeled DNA is plotted taking the following value as the relative distance of sedimentation of each fraction: (volume between the meniscus and the middle of the fraction) minus (one-half of the volume of the loaded sample)/(volume between the meniscus and the middle of the band of δA DNA) minus (one-half of the volume of the loaded sample.

As examples of such plots, the sedimentation patterns of [^{3}H]thymidine and [^{3}H]thymine pulse-labeled T4 DNA are shown in

Fig. 3. In these experiments, the pulse labeling was terminated with KCN and ice, and DNA was extracted by NaOH-EDTA-sodium N-lauroyl sarcosinate treatment. Linear gradients of 5 to 20% alkaline sucrose were made on 80% sucrose cushions. Centrifugation was for 16 hr at 22,500 rpm and at 4°C in an SW25.3 rotor. Recovery of acid-insoluble radioactivity in DNA extraction and in alkaline sucrose gradient sedimentation was 95 to 100% and 85 to 90%, respectively. In the left the radioactive DNA was measured by the routine filter paper method, and in the right the radioactivity in material that was precipitable by cold TCA, was resistant to alkali digestion (0.3 M NaOH at 37°C for 18 hr), and was solubilized by hot TCA treatment (90°C, 30 min) was counted.

Alkaline sucrose gradient sedimentation is also used for preparative purposes. For the isolation of doubly labeled nascent T4 fragments, 5-ml samples of labeled nucleic acid were layered on 30 to 32 ml of 5 to 20% alkaline sucrose gradients and centrifuged at 4°C in a Spinco SW 27 rotor for 19 to 20 hr at 25,000 rpm; 2.8- to 3-ml fractions were collected (Ref. 5, Figs. 3 and 4; Ref. 6, Figs. 7 and 10).

B. Other Methods

Other techniques such as neutral sucrose gradient sedimentation and Sepharose 2B gel filtration are also employed for analysis and isolation of pulse-labeled DNA [1,2,7,8]. An appreciable portion of the pulse-labeled DNA is obtained, without denaturation treatment, as single-stranded fragments after DNA extraction involving deproteinization (e.g., Pronase or phenol treatment). Thus, a portion of the pulse-labeled DNA, not subjected to denaturation, sediments slowly in neutral sucrose, behaves as single-stranded DNA upon hydroxyapatite chromatography and equilibrium density gradient centrifugation, and is susceptible to _E. coli_ exonuclease I [1,2,3,25,26]. This property of the nascent fragments, for which several explanations have been offered [3,25], provides a method for isolation of these molecules.

VI. ANNEALING OF THE NASCENT FRAGMENTS WITH SEPARATED COMPLEMENTARY DNA STRANDS

The complementary strands of T4 and P2 phage DNA are isolated by a method similar to that described by Cohen and Hurwitz [27] for λ DNA. T4D OS or P2 $\underline{vir_1}$ is purified by low- and high-speed centrifugation followed by CsCl density gradient centrifugation [28]. T4D OS containing 200 μg of DNA or P2 $\underline{vir_1}$ containing 100 μg of DNA in CsCl (less than 80 μl) is mixed with 7 ml of 0.1 x SSC-1 mM EDTA-3 mM Tris-HCl (pH 8.5)-0.15% sodium N-lauroyl sarcosinate containing 200 μg of poly(UG), (Miles Laboratories, Inc.). (With some preparations of poly(UG), prior treatment with 0.1 M NaOH at 37°C for several minutes to partially degrade the molecules is necessary.) The mixture is heated at 94° to 98°C for 3 to 4 min and chilled quickly in ice water. After dissolving 9.4 g of CsCl in the sample, it is placed in nitrocellulose tubes and centrifuged in a Beckman 40 or 50Ti rotor for 60 to 70 hr at 30,000 rpm at 10°C. Fractions (45 to 100 μl) are collected from the bottom of the centrifuge tubes, and the optical density of each fraction is measured. Peak fractions of two strands are pooled, and poly(UG) is removed by incubation in 0.3 M KOH at 37°C for 16 to 24 hr followed by dialysis against 2 x SSC and 5 x SSC. The samples are then incubated at 65°C for 4 hr so that "self-annealing" could eliminate the influence of the contaminating complementary strand.

"Fraction H" and "Fraction L" of _B. subtilis_ DNA, which may represent the complementary strands of the chromosomal DNA, are isolated by the method of Rudner, Karkas, and Chargaff [29], which uses a column of methylated albumin kieselguhr (MAK).

A method similar to that described by Denhardt [30] is employed for DNA-DNA hybridization. A solution of the separated DNA strands in 6 x SSC (3 to 5 ml) is passed at a rate of 6 ml/min through a membrane filter (Sartorius, MF 50, 25-mm disk) that has been washed with 6 x SSC and placed on a filter holder. The membrane is washed with 5 ml of 6 x SSC, transferred into a vial, and dried in vacuo at room temperature overnight. It is heated in an oven at 80°C for 3 hr and transferred to a desiccator at room temperature. One milliliter of the preincubation mixture containing 0.5% sodium dodecyl sulfate, 0.2% Ficoll, 0.2% polyvinylpyrrolidone, and 0.2% bovine serum albumin in 3 x SSC is added to the vial. The vial is then placed in a water bath at 65°C for 6 hr. The membrane filter is transferred to a new vial containing labeled DNA sample in 0.6 ml of the preincubation mixture, and incubated for 12 hr in a water bath at 65°C. The membrane filter is washed with 200 ml of 3 mM Tris-HCl (pH 9.35) in a beaker three times. The membrane filter is placed on a filter holder and washed with 50 ml of 3 mM tris-HCl (pH 9.35). It is then placed on the holder upside down and washed with 50 ml of the same buffer. After drying under an infrared lamp, the radioactivity on the membrane filter is counted.

Radioactive DNA is fragmented by heating for 7 min in 0.5 M NaOH at 100°C before use. To ascertain the purity of the separated strands and to provide standards for the experiments with pulse-labeled DNA, model experiments with labeled and unlabeled DNA strands are performed as shown in Figs. 4 and 5, and in Table 1. In some cases a model experiment, as shown in Table 1, is carried out with small amounts of separated labeled strands in the presence of an unlabeled nucleic acid sample, which is prepared in the same way as the radioactive sample to be tested.

TABLE 1

Annealing of Mixtures of the ^{14}C-Labeled T4 W and C Strands with Unlabeled W and C Strands [31]

Labeled strands added W : C[a]	Input annealed with 1 μg of strand, %	
	W strand	C strand
4 : 0	3.5	38.5
3 : 1	11.6	30.6
2 : 2	21.8	20.7
1 : 3	29.8	10.7
0 : 4	40.2	1.0

[a]Total of 925-960 cpm.

An experiment with T4 nascent fragments (8 to 10S) isolated from T4-infected cells is shown in Fig. 6, which indicates equal annealing of the nascent fragments with the separated complementary strands of the phage DNA. Similar experiments with P2 and _B. subtilis_ have demonstrated that the nascent fragments of these organisms anneal predominantly with one of the DNA strands [15, 32,33].

The technique is also used for preparative purposes. Separation of the T4 nascent fragments of both strands was achieved as follows [5,6]. Thirty membrane filters (25-mm disks) loaded with the W strand and the same number of membrane filters loaded with the C strand were prepared by immobilizing 2 μg of the isolated strand to each disk. The two types of membrane filters (W and C) were placed alternately and horizontally in a vial containing 8 ml of solution of the labeled nascent fragments in 3 x SSC containing 0.02% Ficoll, polyvinylpyrrolidone, and bovine serum albumin, and incubated for 12 hr at 65°C. The membrane filters were washed three times in 500 ml of SSC, and the labeled nascent fragments annealed to each phage DNA strand were eluted by dipping each type of membrane filter in 5 ml of 0.1 N NaOH containing 10 mM EDTA at room temperature for 10 min.

VII. TEMPERATURE-SHIFT EXPERIMENTS TO TEST THE FUNCTION OF GENE PRODUCTS IN THE METABOLISM OF THE FRAGMENTS

Insights into the in vivo role of a protein can be gained by studying the immediate metabolic effect of transfer of a temperature-sensitive mutant for that protein from a permissive to a restrictive temperature.

The involvement of polynucleotide ligase in the joining of the nascent fragments has been demonstrated using temperature-sensitive ligase mutants of T4 [3,4,34,35]. The nascent fragments are accumulated upon an upward temperature shift in T4 _ts_ A80- or B20 (gene 30, ligase)-infected cells but not in T4D (wild type)-infected cells (Fig. 7). The accumulated fragments are joined if the temperature is shifted downward (Fig. 8). Involvement of polynucleotide ligase in the joining of the nascent fragments in _E. coli_ has also been shown by similar experiments with temperature-sensitive bacterial ligase mutants [36-38].

The joining of the nascent fragments is also abnormally slow under DNA polymerase I-deficient conditions. This has been demonstrated using a temperature-sensitive mutant, as well as amber mutants of the _pol_A gene [18,38-40]. Figure 9 shows an experiment with _E. coli_ C2107 (_pol_A^{ts}) comparable with the experiment with T4 gene-_30_ mutants presented in Fig. 7. Temperature-shift experiments with P2-infected _E. coli_ C2107 (_pol_A^{ts}) [15] have shown slow joining of P2 nascent fragments under DNA polymerase I-deficient conditions. They have further provided evidence that both strands of P2 are replicated discontinuously, although the discontinuity occurs predominantly in one of the strands (H strand) under normal conditions.

For temperature shifts upwards, the culture is simply poured into a flask being shaken in a water bath of the new temperature when the volume of the culture is 10 ml or less. With larger volumes, the culture is warmed to the new temperature by shaking in a water bath at a temperature higher than that desired before it is placed in the water bath at the new temperature. By these methods, temperature shifts from 20° up to 43°C are completed in 1 min. For temperature shifts downward, the culture is chilled in ice water to the desired temperature. It takes about 10 sec to change the temperature of a 30-ml culture from 43° to 30°C. The quick change of temperature is followed fairly accurately using a sensitive thermistor thermometer.

VIII. ATTACHMENT OF RNA

As noted in the introduction, evidence has been obtained recently that RNA is attached to the 5' end of the nascent fragments. Major lines of evidence for RNA attachment to the *E. coli* fragments are (a) the higher density of the fragments than the bulk DNA [7,9], (b) transfer of ^{32}P from $[\alpha\text{-}^{32}P]$-deoxyribonucleoside triphosphate to ribonucleotides upon hydrolysis of the fragments synthesized by toluene-treated cells with the labeled triphosphates by alkali or pancreatic RNase [8], and (c) creation of a 5'-OH terminus of DNA upon hydrolysis of the fragments with alkali or pancreatic RNase [9]. The methods used have been described in the original papers; they may need improvement and refinement.

REFERENCES

1. K. Sakabe and R. Okazaki, Biochim. Biophys. Acta, 129, 651 (1966).
2. R. Okazaki, T. Okazaki, K. Sakabe, K. Sugimoto, and A. Sugino, Proc. Natl. Acad. Sci. U.S., 59, 598 (1968).
3. R. Okazaki, T. Okazaki, K. Sakabe, K. Sugimoto, R. Kainuma, A. Sugino, and N. Iwatsuki, Cold Spring Harbor Symp. Quant. Biol., 33, 129 (1968).
4. K. Sugimoto, T. Okazaki, and R. Okazaki, Proc. Natl. Acad. Sci. U.S., 60, 1356 (1968).
5. T. Okazaki and R. Okazaki, Proc. Natl. Acad. Sci. U.S., 64, 1242 (1969).
6. A. Sugino and R. Okazaki, J. Mol. Biol., 64, 61 (1972).
7. A. Sugino, S. Hirose, and R. Okazaki, Proc. Natl. Acad. Sci. U.S., 69, 1863 (1972).
8. A. Sugino and R. Okazaki, Proc. Natl. Acad. Sci. U.S., 70, 88 (1973).
9. S. Hirose, R. Okazaki, and F. Tamanoi, J. Mol. Biol., 77, 501 (1973).
10. G. Magnusson, V. Pigiet, E. L. Winnacker, R. Abrams, and P. Reichard, Proc. Natl. Acad. Sci. U.S., 70, 412 (1973).
11. S. Sato, S. Ariake, S. Saito, and T. Sugimura, Biochem. Biophys. Res. Commun., 49, 827 (1972).
12. M. A. Waqar and J. A. Huberman, Biochem. Biophys. Res. Commun., 51, 174 (1973).
13. C. Anagnostopoulos and J. Spizizen, J. Bacteriol., 81, 741 (1961).
14. R. Calendar, B. Lindqvist, G. Sironi, and A. J. Clark, Virology, 40, 72 (1970).
15. Y. Kurosawa and R. Okazaki, J. Mol. Biol., submitted.
16. A. Sugino and R. Okazaki, unpublished observations.

17. H. Manor, D. Goodman, and G. S. Stent, J. Mol. Biol., 39, 1 (1969).

18. R. Okazaki, M. Arisawa, and A. Sugino, Proc. Natl. Acad. Sci. U.S., 68, 2954 (1971).

19. K. I. Berns and C. A. Thomas, Jr., J. Mol. Biol., 11, 476 (1965).

20. E. H. McConkey, in Methods in Enzymology (L. Grossman and K. Moldave, eds.), Vol. 12A, Academic, New York-London, 1967, pp. 620-634.

21. R. Okazaki, in Methods in Enzymology (L. Grossman and K. Moldave, eds.), Vol. 21D, Academic, New York-London, 1971, pp. 296-304.

22. J. D. Friesen, in Methods in Enzymology (L. Grossman and K. Moldave, eds.), Vol. 12B, Academic, New York-London, 1968, pp. 625-635.

23. F. J. Bollum, in Methods in Enzymology (L. Grossman and K. Moldave, eds.), Vol. 12B, Academic, New York-London, 1968, pp. 169-173.

24. R. L. Sinsheimer, in Methods in Enzymology (L. Grossman and K. Moldave, eds.), Vol. 12B, Academic, New York-London, 1968, pp. 850-858.

25. M. Oishi, Proc. Natl. Acad. Sci. U.S., 60, 392 (1968).

26. M. Oishi, Proc. Natl. Acad. Sci. U.S., 60, 691 (1968).

27. S. N. Cohen and J. Hurwitz, Proc. Natl. Acad. Sci. U.S., 57, 1759 (1967).

28. R. E. F. Mathews, Virology, 12, 521 (1960).

29. R. Rudner, J. D. Karkas, and E. Chargaff, Proc. Natl. Acad. Sci. U.S., 60, 630 (1969).

30. D. T. Denhardt, Biochem. Biophys. Res. Commun., 23, 641 (1966).

31. K. Sugimoto, T. Okazaki, Y. Imae, and R. Okazaki, Proc. Natl. Acad. Sci. U.S., 63, 1343 (1969).

32. R. Kainuma-Kuroda and R. Okazaki, J. Mol. Biol., submitted.

33. R. Kainuma and R. Okazaki, J. Japan. Biochem. Soc., 42, 464 (1970).

34. J. Newman and P. Hanawalt, J. Mol. Biol., 35, 639 (1968).

35. J. Hosoda and E. Mathews, Proc. Natl. Acad. Sci. U.S., 61, 997 (1968).

36. C. Pauling and L. Hamm, Proc. Natl. Acad. Sci. U.S., 64, 1195 (1969).

37. E. B. Konrad, P. Modrich, and I. R. Lehman, J. Mol. Biol., 77, 519 (1973).

38. M. M. Gottesman, M. Hicks, and M. Gellert, J. Mol. Biol., 77, 531 (1973).

39. P. L. Kuempel and G. E. Veomett, Biochem. Biophys. Res. Commun., 41, 973 (1970).

40. K. Shinozaki, Y. Kurosawa, and R. Okazaki, in preparation.

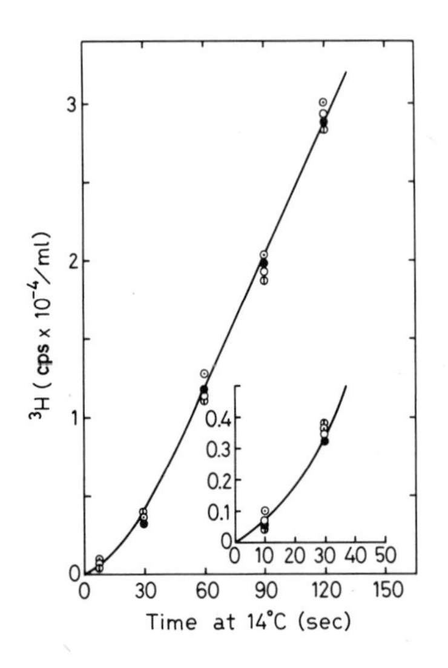

FIG. 1. Kinetics of [^{3}H]thymidine incorporation into *E. coli* DNA studied with various stopping methods [16]. *E. coli* 15TAMT was grown at 30°C in Medium A supplemented with 0.5% Casamino Acids, 0.01% L-tryptophan, and 40-μM thymine to about 9 x 10^8 cells/ml, transferred to 14°C, and stirred for 90 min. At time zero, [^{3}H]thymidine (18 Ci/mmole) was added to a final concentration of 0.2 μM and, at the times indicated, 2-ml samples were added to the following mixtures: (a) 2 g of crushed ice and 0.4 ml of 0.2 M KCN (o); (b) 8 ml of acetone at -20°C (⊙); (c) 2 ml of "ethanol-phenol" (⊕); or (d) 2 ml of 10% trichloracetic acid (TCA) at 0°C (●). In every case, cells were dissolved in 0.2 M NaOH, then precipitated and washed with cold TCA; the radioactivity was then counted.

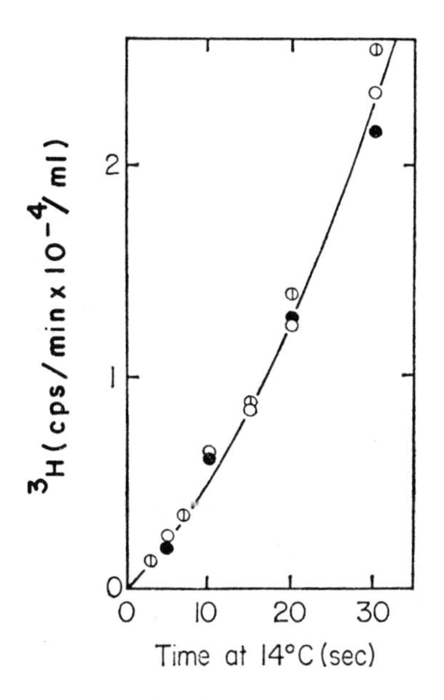

FIG. 2. Kinetics of [^{3}H]thymidine incorporation into T4 DNA studied with various stopping methods [6]. A culture (75 ml) of *E. coli* B infected with T4D at 20°C was divided into three portions and chilled to 14°C at 70 min of infection. After 10 min at 14°C, [^{3}H]thymidine (19.2 Ci/mmole) was added to each portion to a concentration of 0.2 μM. At the times indicated, 4-ml samples of each were taken into three mixtures: (a) 3 g of crushed ice and 0.6 ml of 0.2 M KCN (o); (b) 3 ml of "ethanol-phenol" (⦶); or (c) 0.45 ml of 50% TCA at 0°C (●). Cells were collected by centrifugation. In (b), cells were washed once with 10 ml of 75% ethanol-18 mM sodium acetate-2% phenol-0.2 μM thymidine. The radioactivity was measured as in Fig. 1.

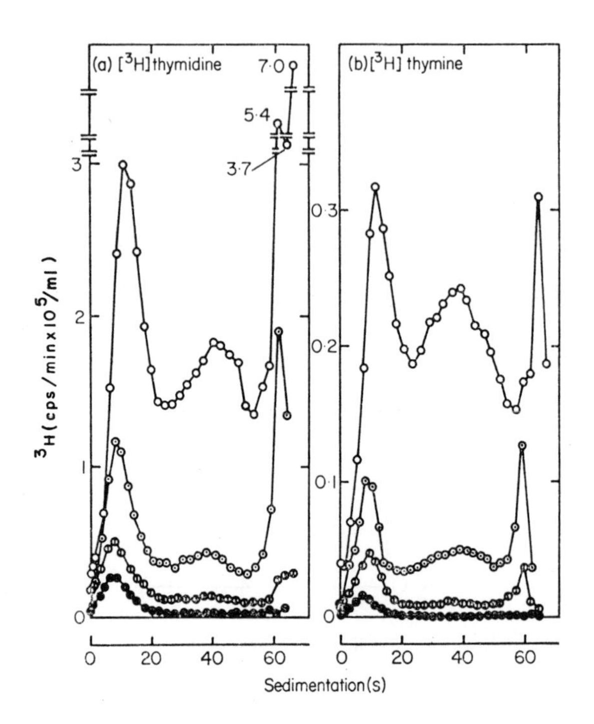

FIG. 3. Alkaline sucrose gradient sedimentation of pulse-labeled DNA from T4-infected E. coli [6]. Left (a) A culture of E. coli B (5×10^8 cells/ml in Medium A) was infected with T4D (wild type) at a multiplicity of 10 at 20 C and chilled to 8 C, 10-ml portions of the culture were pulsed labeled with 0.5 μM [^{3}H]thymidine (21 Ci/mmole) for the times indicated. Right (b) A culture of E. coli JG116 (6×10^8 cells/ml in Medium A supplemented with 16 μM thymine) was infected with T4 td8 at a multiplicity of 10 at 20 C and chilled to 8 C at 70 min of infection. After 10 min at 8 C, 10-ml portions of culture were pulse labeled by the addition of 1 μM [^{3}H]thymine (20.6 Ci/mmole). —●—●— , 5 sec; —◐—◐— , 30 sec; —◎—◎— , 60 sec; —○—○— , 300 sec.

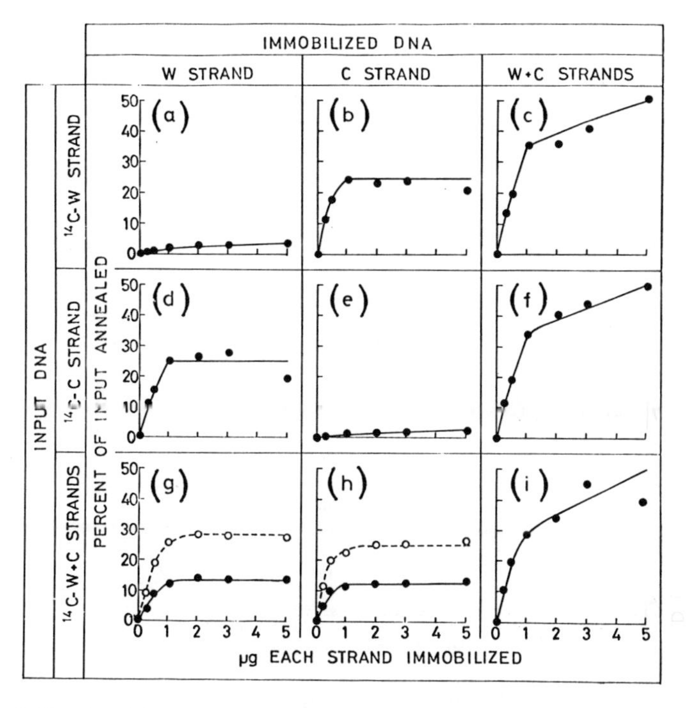

FIG. 4. Model annealing experiments with labeled and unlabeled W and C strands of T4 DNA: Annealing of the labeled strand(s) as a function of the amount of unlabeled strand(s) immobilized [31]. Each partially degraded labeled strand [0.16 μg (750 cpm)] was annealed with the unlabeled strand(s) as indicated. "W + C strands" denotes an equal mixture of the W and C strands. —●— : Percent annealing with respect to the total labeled strand(s) added. --○--: Percent annealing with respect to the labeled strand complementary to the immobilized unlabeled strand.

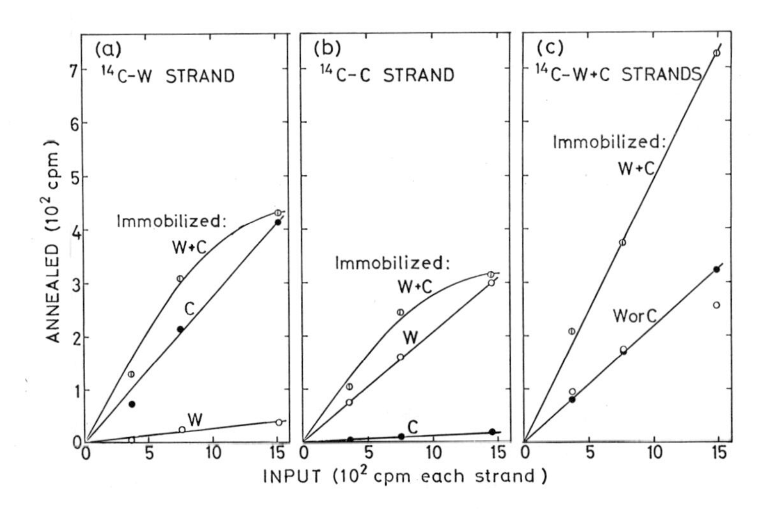

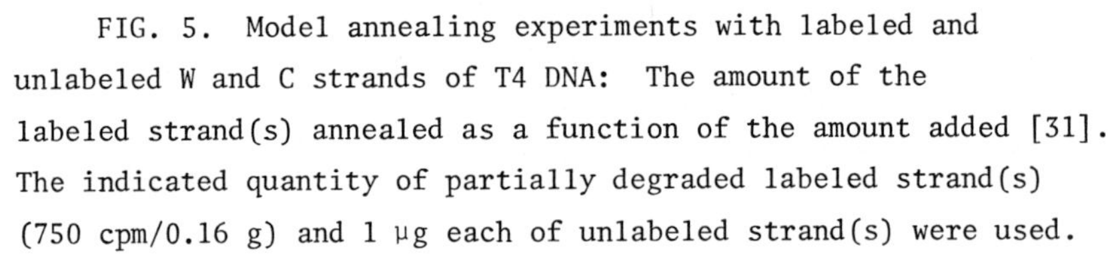

FIG. 5. Model annealing experiments with labeled and unlabeled W and C strands of T4 DNA: The amount of the labeled strand(s) annealed as a function of the amount added [31]. The indicated quantity of partially degraded labeled strand(s) (750 cpm/0.16 g) and 1 μg each of unlabeled strand(s) were used.

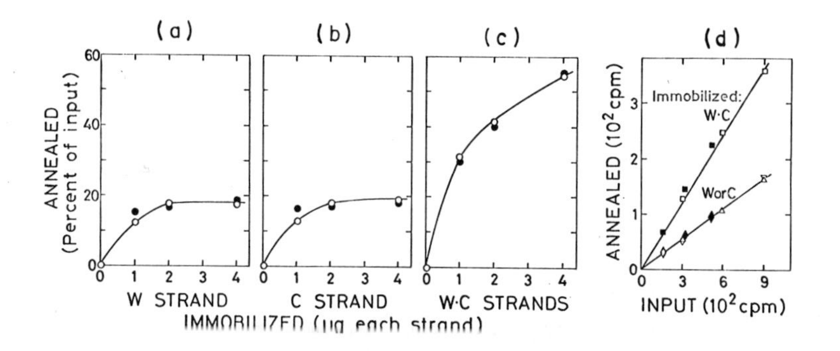

FIG. 6. Annealing with the W and C strands of the nascent fragments isolated by alkaline sucrose gradient sedimentation from T4D-infected cells pulse labeled with [^{3}H]thymidine for 30 sec at 20°C at 70 min of infection [31]. In (a), (b), and (c), 590 cpm of the ^{3}H-labeled fragments (o) or 365 cpm of partially degraded ^{14}C-labeled T4 DNA (•) were annealed with the unlabeled strand(s) as indicated. In (d), the indicated amount of the ^{3}H-labeled fragments (o, △, ▽) or partially degraded ^{14}C-labeled T4 DNA (•, ▲, ▼) was annealed with a fixed amount (2 μg each) of unlabeled strand(s). "W·C strands" denotes unfractionated T4 DNA.

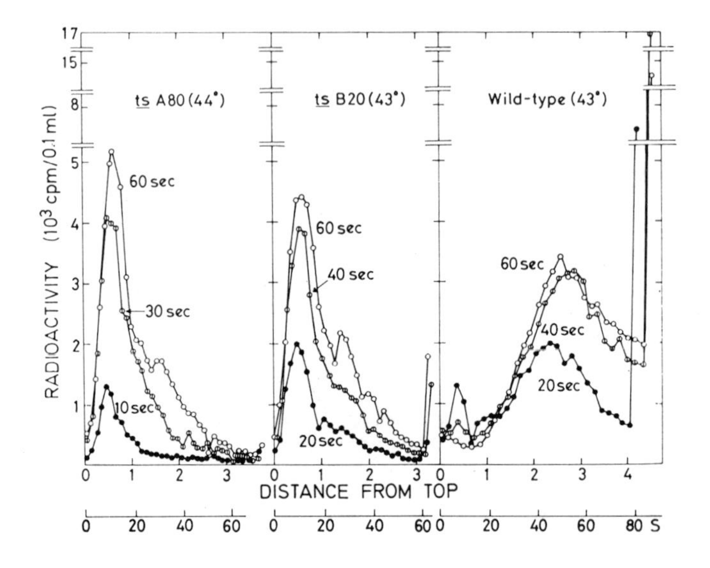

FIG. 7. Alkaline sucrose gradient sedimentation of DNA from gene 30 ts mutant- and wild-type T4-infected cells pulse-labeled at high temperature [4]. Cells were infected with the indicated strain of T4 and incubated at 20°C for 70 min before the temperature was elevated to 43 or 44°C. The cells were pulse labeled with 10^{-7} M [^{3}H]thymidine (15 Ci/mmole) for the indicated time beginning at 2 min after the temperature shift. DNA was extracted by NaOH-EDTA treatment and sedimented through 5 to 20% linear sucrose gradients containing 0.1 M NaOH, 0.9 M NaCl, and 1 mM EDTA in a Spinco SW 25.3 rotor for 13 to 16 hr at 22,500 rpm and 4°C. The distance from the top is relative to that of infective δA DNA (19S) used as an internal reference.

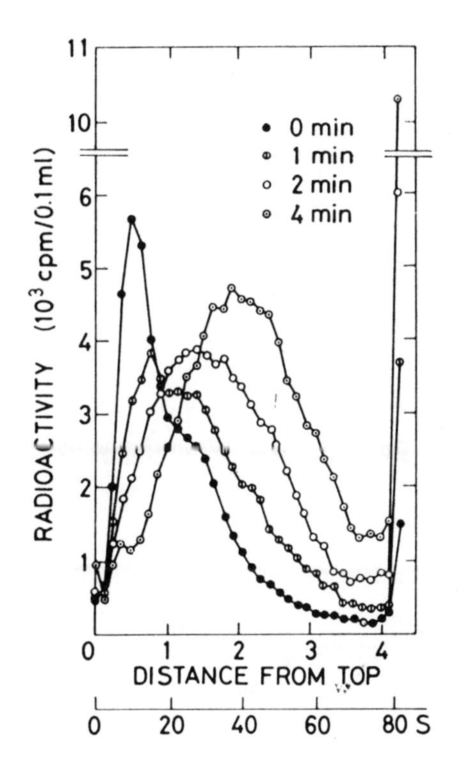

FIG. 8. Effect of incubation at low temperature subsequent to pulse labeling at high temperature on the sedimentation pattern of radioactive DNA from T4 ts B20-infected cells [4]. Cells were pulse labeled for 60 sec with [^{3}H]thymidine at 43°C as in Fig. 7 and then incubated at 30°C for the indicated time. DNA extraction and alkaline sucrose gradient secimentation were carried out as in Fig. 7.

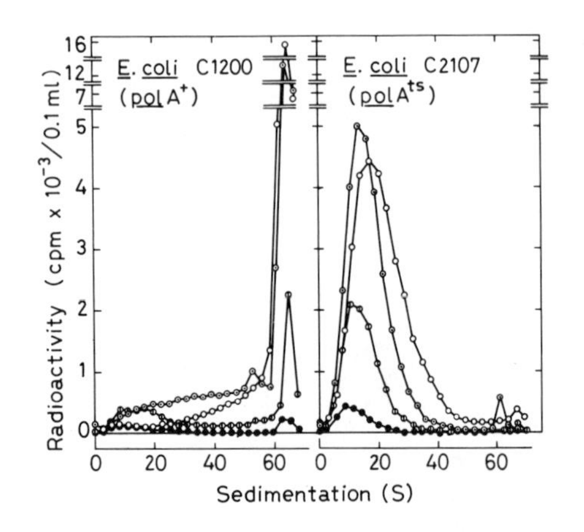

FIG. 9. Alkaline sucrose gradient sedimentation of DNA from a temperature-sensitive E. coli polA mutant (C2107) and wild-type E. coli (C1200) pulse labeled at high temperature [40]. Cells were grown at 30°C in Medium A supplemented with 0.5% Casamino Acids to a titer of 4 x 10^8 cells/ml, transferred to 43°C, and 2 min later pulse labeled with 10^{-7} M [^{3}H]thymidine for the indicated times. The DNA was extracted by NaOH-EDTA-sodium N-lauroyl sarcosinate treatment and sedimented through a 5 to 20% linear sucrose gradient containing 0.1 M NaOH, 0.9 M NaCl, and 1 mM EDTA in a Spinco SW 41 rotor for 10.5 hr at 27,000 rpm at 4°C. The distance from the top is relative to that of ^{14}C-labeled δA phage DNA (19S) used as an internal marker. —o—o— 120 sec; —⊙—⊙— 60 sec; —◐—◐— 20 sec; —●—●— 7 sec.

Chapter 2

APPLICATION OF ELECTRON MICROSCOPIC DENATURATION MAPPING TO THE STUDY OF DNA REPLICATION

Dhruba K. Chattoraj* and R. B. Inman

Biophysics Laboratory and Biochemistry Department
University of Wisconsin
Madison, Wisconsin

---

*Present address: Institute of Molecular Biology, University of Oregon, Eugene, Oregon.

I. THE RATIONALE OF THE TECHNIQUE

Electron microscopy of DNA has been widely used in the study of DNA replication. Direct visualization of the replicating intermediates has proven to be very useful in our understanding of the mechanism of DNA replication. The application of the technique of denaturation mapping has provided for the first time a straightforward and precise means of physically mapping the position of the replicating forks in certain phage DNA molecules.

The principles involved in denaturation mapping exploit certain properties of the helix-random coil transition exhibited by the DNA double helix. It has long been known that the transition temperature (T_m) of a DNA sample is a linear function of the base composition (G + C/A + T) of that DNA; this property has, therefore, provided a reliable parameter for the determination of DNA base composition [1]. A variety of reagents have been reported that destabilize the DNA helix; these include alkali [2], formamide [3], formaldehyde [4], and the DNA unwinding proteins of *E. coli* [5] or T4 phage [6]. These are discussed below in connection with studies involving electron microscopic denaturation mapping.

As might be expected, the helical stability is again a reflection of base composition at the intramolecular level. When a DNA solution is partially denatured then, in individual molecules, the A + T richest sequences are the first to denature; these can be detected in the electron microscope as dissociated single-stranded regions in otherwise native DNA. As the degree of denaturation is increased, the segments that next denature will contain decreasing amounts of A + T and, finally, at high degrees of denaturation the only double-stranded helical regions that remain will be those richest in G + C [7,8]. The preferential opening of dA·dT rich segments along a DNA molecule forms the basis of the partial denaturation mapping technique. We will now restrict our discussion to the partial denaturation of a homogenous DNA sample (where the individual molecules are of identical length and base sequence). In this case, if the DNA molecule has sufficient base compositional heterogeneity along its length, the denatured regions occur at identical positions in all the molecules and they can, therefore, be used as a frame of reference for the study of any other point of interest on the molecule. In several instances, the denaturation maps (which are simply linear representations of the size and position of denatured segments) were found to be a unique property of the molecule (examples are shown in Fig. 1). It can be seen immediatly that the position of the denatured sites is highly asymmetric with respect to the two halves of the molecule. The ends of the molecules are arbitrarily aligned in such a way that the right halves contain the majority of the denatured sites. The maps shown in Fig. 1 allow the following simple but unambiguous conclusions; first, DNA molecules of λ, P2, 186 and P4 phages can easily be recognized by their denaturation pattern and second, the left and right ends of each molecule can be defined on the basis of its denaturation map. This information can be exploited to answer many important questions relating to the structure and function of DNA. Various applications of this physical mapping technique have recently been reported elsewhere [9,10]. In the present chapter, a description of the technique and its application to the study of DNA replication is discussed and is restricted mainly to the work done in this laboratory.

II. METHODS

A. Partial Denaturation of DNA

1. Thermal

A solution containing 0.1 M NaCl, 0.0067 M KH_2PO_4, 0.0034 M EDTA, 10% HCHO [(Matheson, Coleman and Bell); the concentrated formaldehyde solution is first neutralized and heated for 10 min at 100°C], and DNA (OD_{260}=0.005-0.01) at a final pH between 6.5 and 7.5 is heated at various temperatures in a constant temperature bath for 10 min. The sample is quickly cooled in an ice bath and prepared for electron microscopy (see Section II,B,2).

The degree of denaturation is determined by the temperature, time of heating, and the G + C content of the sample. For instance, for a heating period of 10 min, a DNA with an average G + C content of 50% begins to denature at 48°C and is completely denatured at 58°C [7]. The differential denaturation of the dA·dT and dG·dC rich segments (which is the aim of partial denaturation mapping) is also sensitive to the ionic strength; one would expect greater discrimination between A + T and G + C rich sequences when the salt concentration is lowered. Thus, depending on the DNA and the salt concentration, time and temperature must be chosen to obtain the desired degree of denaturation. The exact condition can be found by inspection of the published thermal denaturation maps of the following DNAs; bacteriophages--λ[7,8], λb2 [11], P2 [12], T7 [13], and Rhizobium phage [14]; animal viruses--human papilloma [15], polyoma [16,17], adenovirus type 2 [18], adenovirus type 12 [19], and SV40 [20]; and mitochondria--rat liver [21].

2. High pH

The buffer for high pH denaturation is made by mixing 0.32 ml of 1 M Na_2CO_3, 0.4 ml of 0.126 M Na_2EDTA, and 4 ml of 40% HCHO (Matheson, Coleman and Bell). The pH of the solution should be about 10.0. This is then adjusted to the denaturation pH by adding 5 M NaOH solution. Three microliters of this buffer are put into a small conical base glass bottle (volume of the conical segment is 0.1 ml, Kontes Glass Co., N.J.) immediately after rinsing the bottle with the same pH buffer. Seven microliters of the DNA solution (OD_{260}=0.01-0.03) in 0.02 M NaCl + 0.005 M Na_2EDTA (pH 7.5) are added and kept at 23°C for 10 min. The mixture is chilled before preparing for electron microscopy (see Section II,B,2). This solution can be neutralized prior to preparation for electron microscopy [22]; however, we have obtained somewhat better results when neutralization is omitted.

The high pH buffer is made up not more than an hour before use. Under the above conditions, bacteriophage DNA with approximately 50% G + C content (Fig. 1) starts to denature at about pH 11.0, and denaturation is almost complete at pH 11.4. As with thermal denaturation, small changes in pH, time, temperature, and ionic strength can effect the final degree of denaturation. The best conditions can be estimated by inspection of published high pH denaturation maps from bacteriophages-- λ[22,23], P2 [24,25], 186 [26,27], P4 (Fig. 1), T7 [13]; animal viruses-- SV40 [28], and CELO [29]; ribosomal--*Xenopus laevis* [30,31] and *Xenopus mulleri* [31]; and 5S DNA of *Xenopus laevis* [32].

3. Formamide

With minor changes in the method described above (Section II, A,2), partial denaturation can be achieved with formamide in the absence of HCHO. Three microliters of a buffer solution containing a 0.068 M Na_2CO_3 + 0.0107 M Na_2EDTA (pH 10.4) is mixed with 7 μl of DNA (OD_{260} = 0.03 - 0.09) in 0.02 M NaCl + 0.005 M Na_2EDTA (pH 7.5) and then adjusted to the desired formamide concentration. The incubation is carried out for 10 min at 23°C and the mixture is chilled for 5 min before cytochrome is added to 0.01% (as in Section II,B,2). The final formamide concentration was varied between 80 and 90%. A somewhat different procedure for formamide denaturation has been reported from other laboratories [33,34].

Partial denaturation by formamide has the advantage that the maps presumably represent a denatured state at thermodynamic equilibrium (this is certainly not the case with heat and high pH denaturation, which involve the use of HCHO). One could therefore hope to relate directly the denaturation map data to actual A + T composition and melting curves. Formamide denaturation, however, produces in our hands a significant number of molecules in which denaturation proceeds from the ends of molecules or by expansion of denatured sites (such an "end effect" is not observed when denaturation is carried out by heat or high pH in the presence of HCHO). This observation rests on a single experiment involving 30 DNA molecules from λ phage and needs to be investigated further.

4. Gene 32 Protein

The T4 bacteriophage gene 32 protein-induced partial denaturation of double-stranded DNA has recently been reported [6]. DNA [OD_{260} = 0.2 in 0.01 M potassium phosphate (pH 7.0), 0.001 M Na_3EDTA] is incubated for 10 min at 37°C with 100 μg gene 32 protein/ml. Gluteraldehyde is added (to a final concentration of 0.01 M) and the incubation is continued for another 10 min. This concentration of gluteraldehyde is chosen to fix the gene

32 protein to the single strands of DNA and thereby stabilize the protein-DNA complex during the subsequent preparative procedures. After tenfold dilution in a buffer containing 0.1 M Tris (pH 8.5), 30% formamide, and 0.01% cytochrome-*c*, the mixture is spread onto a solution containing 10% formamide and 0.01 M Tris (pH 8.5). The cytochrome film is picked up on parlodion-coated copper grids, stained with uranyl acetate for 20 sec (0.005 M uranyl acetate + 0.05 M HCl in 95% ethanol, diluted 1000-fold into 90% ethanol just before use), rinsed in 90% ethanol, dried from isopentane, and finally shadowed with platinum.

B. Specimen Preparation for Electron Microscopy

The basic protein monolayer technique for preparing double-stranded DNA for electron microscopy [35] has been modified so that single strands can be clearly visualized. The single-stranded regions should take up an extended configuration, but the double-stranded DNA should remain stable. The presence of formamide during the spreading procedure yields good extension of single strands and additionally greatly improves their contrast [36,37]. The particular method routinely used in this laboratory [22] is described below.

1. Preparation of Carbon-Coated Mica Disks

Mica disks (3/8 in diam) are punched out from freshly cleaved mica sheets (Ladd Research Industries, or Ted Pella Company) with a blanking punch (Di-Acro Company, Lake City, Minnesota) held in a drill press. Prior to punching, the mica is placed, freshly-cleaved side down, on a piece of clean paper and taped along two edges. The paper protects the mica during and after the punching operation. The mica disks are then cleaned with a stream of air and placed in a vacuum, and a carbon film is evaporated onto the freshly cleaved side. Presumably, a worthwhile precaution at this stage is to minimize oil-vapor contamination during evacuation. The thickness of the carbon film must be determined by trial and error; it should be as thin as possible, but thick enough to survive the various manipulations that follow. A 1/16-in outer annulus of carbon film is now scratched away from each disk. This step produces a hydrophilic ring around the carbon film that is needed when the DNA sample is later attached to this film. The scratching operation can be easily performed by a rotating annular brush and a vacuum hold-down device for the disk. Large numbers of these carbon-coated disks can be prepared at one time and then stored at about 10% humidity. Our impression is that these disks work best after they have "aged" for a week or more, but that after about 4 months they may no longer be satisfactory.

2. Spreading of DNA Solution on a Water Drop

The partially denatured DNA solution (see Section II,A,1 and 2) is mixed with an equal volume of cold (4°-10°C) formamide and cytochrome-*c* crystallized twice and lyophilized, Calbiochem) is added to a final concentration of 0.01%. After standing for 5 min at 23°C, the mixture is spread on a 1.2 ml drop of double-distilled water (placed in an indentation on a Teflon block). The Teflon block (15 cm x 15 cm x 0.2 cm) is bonded to an aluminum support which in turn is mounted on leveling screws. Twenty-four hemispherical indentations are milled into the top surface of the Teflon, each being 1.9 cm in diameter and about 0.1 cm deep. Before use, the Teflon block is cleaned by brushing the surface with detergent and rinsing well with distilled water.

By using a 0.5 cm i.d. capillary pipet we transferred 0.005 ml of the spreading solution to a clean glass rod (0.3 cm diam and drawn to a fine, but rounded tip at one end), as indicated in Fig. 2. The glass rod should be kept wet to assure that the spreading solution runs down the rod evenly. The rod is now carefully removed from the side of the drop. The surface of the film can be compressed by withdrawing 0.1 ml of water from within the drop, using a syringe with a fine needle. A DNA sample can then be picked up by touching a carbon film (evaporated onto a mica surface as described in Section II,B,1) to the surface of the drop. The mica disk, with adhering carbon film and drop of liquid, is now immersed and washed in ethyl alcohol and dried in a stream of warm, dry nitrogen gas. Alternatively, the disk can simply be dipped into alcohol and air dried if the ambient relative humidity is less than about 40%. Usually, the DNA solution is spread onto three or four separate drops and each is sampled once with a carbon-coated mica disk. The disks containing the DNA are then rotary shadowed with platinum (3 cm of 0.008 in. diam. platinum wire wound around a tungsten filament, 0.035 in. diam.). The distance from the filament to the center of the rotating table is 10 cm and the height from the table to the filament is 1 cm. After shadowing, the carbon film (containing the cytochrome film and the DNA) is floated off the mica onto a clean water surface, picked up on a specimen grid, and is now ready to be examined in the electron microscope.

The above method has several advantages. Very small amounts of DNA are required, and the carbon film and DNA sample are supported by mica throughout the procedure, which prevents breakage of the carbon film and ensures a flat surface during shadowing. The method is relatively fast and easily allows one to prepare many samples at the same time.

3. Electron Microscopy

Platinum shadowed samples can be examined by normal bright-field electron microscopy at original magnifications between X4000 and X7000. The exact magnification for photography is determined by the length of the molecule, the type of camera, and the method used to measure the micrographs. The preparative method described above yields single-stranded regions which are thinner and have lower contrast than double-stranded DNA; thus, the double and single strands can be clearly distinguished. It has been our experience that unknown effects can produce drying artifacts from time to time. One should guard against basing conclusions on measurements made from molecules that appear to have been aligned during the preparative procedure. In such molecules, the observed length may not be meaningful and the single-stranded denatured regions may not be well resolved.

C. Measurement, Computation, and Display of Denaturation Maps

Denaturation map data can be obtained from the electron micrographs in a variety of ways. One of the simplest methods consists of projecting the micrographs onto a paper screen, tracing the molecules with a pencil, and finally measuring the tracings containing the marked positions of the denatured sites with a map measure. The map measure should have a swiveling handle which acts as a castor to facilitate the measurement (Keuffel and Esser Co., stock item 620300). The denaturation maps can then be constructed by hand or the data can be read into a computer for plotting. Several of the more sophisticated desk-top calculators can be interfaced to digitizing devices that generate x and y coordinates as a handheld pen is moved over a projected image of the micrograph. Three such systems have been tested in this laboratory.

First, a Hewlett Packard 9820 (or 9810) calculator and 9864 A digitizer can be used to digitize and manipulate data from electron micrographs. A Hewlett-Packard 9862-A plotter can also be interfaced to this system to allow the immediate plot output of denaturation maps. The disadvantage of the system is that the digitizer is opaque and, therefore, the micrograph cannot be rear-projected onto the working surface of the digitizer.

Second, a Hewlett-Packard 9820 (or 9810) calculator can be interfaced to a Numonics Corporation Graphics Calculator. The advantage of this system is that the Numonic digitizer can work on any surface and rear projection is, therefore, possible. (Similarly, the digitizer could be used directly over a television screen from an image intensifier-electron microscope system.) Again, a Hewlett-Packard 9862-A plotter would allow the immediate plot-out of denaturation maps once the digitizer has been traced across the molecule and the site positions recorded. Finally, the Numonics Corporation Graphics Calculator can be obtained with an option that allows lengths to be calculated directly by the digitizer rather than by an additional desk-top calculator. These lengths could then be out-put to a computer via a teletype or the data can be treated manually. The first two systems (involving a programmable calculator) allow various types of smoothing manipulations to be performed on the x and y digitizer output before the actual calculation of length. The latter system does not allow such calculations.

Once the molecules have been measured, then two types of correction should be applied. First, on partial denaturation there is a small change in length associated with the conversion from double- to single-stranded DNA [12,18,22]. Single-strand lengths should therefore be corrected to correspond to native lengths. The correction factor can be found by comparing the length of single- and double-stranded ϕX174 circles. For the high pH buffer system described above (Section II,A,2), we have observed that single-stranded ϕX174 is 4.6% longer than the corresponding double-stranded molecule at pH 10.0. After the above correction has been made to each single-stranded region, then the data for each molecule should be normalized to either unit length or to the average length (in micrometers) of all molecules in the experiment. It is advisable to use a computer to manipulate and plot-out the large amount of data involved in denaturation mapping. Computer programs have been written to handle most problems associated with this technique and these programs are available from the authors.

III. APPLICATIONS

The strategy in the application of the electron microscopic denaturation mapping technique to the investigation of DNA replication is to use the denaturation map as a frame of reference. Once a frame of reference has been established, then the location of replication branch points, single-stranded regions, or concatemer joints can be investigated.

A. Origin and Direction of Replication

The strategies to be followed depend somewhat on the particular structure of the replicating intermediate as they are described below.

1. Single Branched Circles

In these molecules the replicating intermediate is circular with an extra linear piece of DNA connected to the circle at one point [Figs. 3(a) and 4]. The origin and direction of replication has been determined for these structures in the case of *E. coli* bacteriophages λ[23], P2 [24], 186 [27]. The mature forms of these phage DNAs are linear and have unique denaturation maps as shown in Fig. 1. Upon infection, the linear DNA becomes circular by the joining of its cohesive ends. This does not alter the denaturation map since the base sequence remains the same. Circular denaturation maps can therefore be compared with the mature linear maps. By alignment of characteristic denatured regions, it is usually possible to determine the position on the circle, which corresponds to the mature ends of the molecule. Such a position can then be used as a frame of reference for the study of branch-point position. Figure 5(a) shows examples of circular denaturation maps of 186 DNA artificially broken, for display purposes, at a point corresponding to the position of the mature ends. Comparison with linear 186 maps [Fig. 6(b)] shows that this method can be used to locate the mature ends with reasonable precision.

The relative position of the branch points can now be investigated. In the cases of λ[23], P2 [24], and 186 [27] replicating DNA, this position was found to be randomly distributed along the molecule. Next, the free ends of the branches can be mapped by matching the denaturation map of the branched segment with the artifically broken circle to obtain good alignment of denatured sites between the branch and the corresponding section of the circle (of equal length), either to the left or to the right of the branch point. In the case of 186 DNA, we found that in 30 out of 39 molecules photographed, the branches had to be aligned to the left of the branch point for the best agreement between denatured sites, and in these cases the free end of the branch occurred at a unique place on the circular map. These results are presented in Fig. 7, where the random position of the branch points (short vertical lines) and the unique position of the free ends (short vertical arrow) can be seen. We interpret the position of the free end as the starting point of replication. This means that the branch point is the growing point and that replication proceeds from left to right with respect to the denaturation map of mature 186 DNA. In nine other denatured branched circles, the free ends of the branches did not map at a fixed position (last nine molecules in Fig. 7). All the circles, however, had a recognizable 186 DNA denaturation map. This latter result could mean that there can be several different origins of replication in a small fraction of the molecules. However, shear breakage of the branches during specimen preparation would also give rise to apparent random origins of replication by our mapping procedure.

2. Double Branched Circles

These molecules have the appearance of the Greek letter theta as shown in Figs. 3(b) and 8, where three segments are connected to each other at two branch points. Two of the three segments are equal in length and correspond to the daughter segments. The length of one daughter segment when added to the length of the third segment equals the unit length of the genome. Bacteriophage λ replicates primarily as a double-branched circle in the first round of replication [23], and the determination of its origin and direction of replication is described here.

First, the position of the mature ends is determined by comparing the denaturation map of the circle with that of the mature map (as already described in Section III,A,1). The circular map in this case, however, has to include one of the daughter segments (as expected, both daughter segments show similar denaturation patterns, Fig. 8). Once the mature ends have been located, the branch-point position can then be found. In contrast to single-branched circles, both branch points in the present structure appear to move away from a fixed origin and, thus, indicate bidirectional replication [23]. If the assumption is made that bidirectional replication initiates at a single unique point and progresses continuously at the same rate in both directions, then the origin of replication is situated 14.4 $\pm$ 1.3 μm from the left mature end (17.7 $\pm$ 7.4% from the right end). This value is found from the average of the midpoints of daughter segments in double-branched molecules. However, there may be a complication to this simple interpretation because the spread of midpoints about the average value is considerably more than would be expected from experimental error. Similarly, when the midpoint positions are plotted against the degree of replication (as deduced from the length of daughter segments in each molecule), there appears to be a dependence of midpoint position on the degree of replication [Fig. 9(a)]. Possible explanations for this have already been discussed [23,38]. If the best fit line shown in Fig. 9(a) is accepted as meaningful (and we wish again to note that the dispersion of the data is large), then possibly the origin is located 13.5 μm from the left end (22.9% from the right end).

In addition to the double-branched circles with two growing points, about 30% of the replicating molecules replicated unidirectionally as a single-branched circle, either to the left or to the right. The origin can also be determined from these molecules. Figure 9(b) shows a plot of the growing point position at various degrees of replication. The best fit lines for these data yield an origin at 14.3 $\pm$ 0.4 μm in good agreement with the average midpoint position (14.4 $\pm$ 1.3 μm) discussed above.

The use of mature ends as a reference point for the location of branch positions is not obligatory. Some types of DNA replicate as double-branched circles but do not have a linear form in their life cycle. If these DNAs show unique denaturation maps, then the investigation of the origin and the direction of replication can still be carried out; for instance, in the case of plasmid DNA NR-1 (also called R-222), denaturation mapping can still serve as a frame of reference on the circular DNA molecules [51]. Restriction endonucleases can also artificially create linear forms from circles, and these ends can provide a point of reference. In this manner, the origin and direction of replication of the animal virus SV40 was determined [39]. Mature closed-circular SV40 DNA was cleaved by restriction endonuclease R_1, and denaturation mapping on the resulting linear molecules showed that the cleavage occured at a unique site [28]. The double-branched circular molecules were cleaved by R_1 to linear forms [Fig. 3(c)] and the branch points measured with respect to the cleavage site. The origin of replication was then determined [39] by following the method discussed above for λ DNA. The use of restriction endonuclease has the advantage that denaturation mapping was needed only to characterize the cleavage site.

3. Linear Replicating Molecules

In the case of bacteriophage T7, the replicating intermediate is a branched, linear molecule [33]. At low degrees of replication, double-branched rods were observed [Fig. 3(c)]. Thus, replication started at an internal position and denaturation mapping showed that the origin was confined to a unique segment of the molecule. Furthermore, as expected, replication proceeded bidirectionally from the origin on these linear molecules [40].

B. Characterization of DNA Concatemers

During late stages in the latent period of λ phage infection, a new form of intracellular λ DNA can be isolated that is known to be longer than mature molecules. It is believed that these long linear pieces of DNA are precursors for the monomeric mature units that are packaged in the phage head. Denaturation mapping has been used to test this notion [41,42]. The long molecules showed a repeating λ-like pattern of denatured sites. The length of the repeating unit and the distribution of denatured sites within the repeating unit were similar to those found for phage λ DNA. It was concluded that the long molecules were head-to-tail concatemers of λ DNA.

Partial denaturation was achieved at high pH (Section II,A,2). The comparison of the distribution of denatured sites with that of the mature λ DNA was complicated by the fact that the polymeric DNA was quite heterogeneous in length and had no fixed start or end points. Electron microscopic measurements were done only on those fractions of the polymeric DNA whose average length was about twice as long as λ. In order to test that the linear polymers consisted of λ monomers, joined head-to-tail, a gauge was constructed showing the expected position of denatured sites in three λ monomers joined head-to-tail [Fig. 10(b)]. The gauge was reproduced on a paper strip and compared with each test molecule by moving the gauge left or right or by reversing it. The position of best fit was judged by eye and then the λ concatemers were aligned accordingly [an example of such an alignment can be seen in Fig. 10(a)]. It is apparent that the test molecules have a λ-like base sequence and the head-to-tail sequence is repeated at least twice. Thus, from knowledge of the denaturation map of the monomer units, it is possible to recognize and characterize DNA concatemers.

C. D-Loops of Intracellular DNA

In mitochondrial DNA from mouse L cells, a novel form of DNA has recently been described. This form has a displacement loop (D-loop) which is a short, single-stranded DNA region spanning an otherwise double-stranded segment [Fig. 3(d)]. The D-loop molecules were characterized as the first discrete stage of DNA replication [43,44]. A similar intracellular form of λ DNA was also seen in this laboratory when the infecting λ phage is defective in genes necessary for DNA replication [45]. An example is shown in Fig. 11. In this case, however, there was often more than one D-loop per circle. The D-loop positions, with respect to the mature ends of the molecule, were determined by denaturation mapping. The normal denaturation procedure was not used since D-loops could be lost during partial denaturation. The structure is expected to be protected if the hydrogen-bonded, newly synthesized strand is cross-linked to the parental strand prior to denaturation. The intracellular DNA was, therefore incubated in the presence of 5% HCHO for 4 to 24 hr at room temperature and then dialyzed at 4°C against 0.02 M NaCl + 0.005 M Na_2EDTA (pH 7.5) containing 1% HCHO. The sample was then partially denatured as described in Section II,A,2. The denaturation map of the cross-linked molecules showed many smaller than usual denatured sites (indicating that cross-linking was effective), but the overall denaturation pattern remained unchanged and permitted location of the position of the mature ends on the circular molecules. As can be seen in Fig. 12(a), D-loops occur in a large number of different positions. They are, however, significantly infrequent in the region from 8.0 to 10.5 μm, which corresponds to the A + T richest segment of the molecule [Fig. 12(c)]. Absence of D-loops at this position may be an artifact since this is exactly the location from which

D-loops would disappear during partial denaturation unless there was sufficient cross-linking. We cannot be sure of this explanation, however, since the frequency of cross-links was not established in these experiments. Unlike mitochondrial DNA [43,44], the function (if any) of D-loops in λ DNA is still not understood [45].

D. Base Composition of the Origin of Replication

If denaturation mapping is used to determine the origin of replication, we can, at the same time, learn something about the base composition at the origin. As is discussed in Section IV, the degree of denaturation used for replication studies should be minimal, but just high enough to allow an accurate estimate of the position of mature ends. However, from maps at higher degrees of denaturation of mature phage DNA, we can determine if the origin of replication is situated at an A + T- or a G + C-rich segment. On the basis of experiments with several different phages, it appears that the origin is located at a relatively G + C rich position. In the case of 186 DNA, it was found that at an average degree of denaturation of 58%, the position of the origin (situated at 10.5 μm) remained undenatured [Fig. 6(d) and (h)]. However, at 81% denaturation, this position is no longer stable [26]. In λ [8,23], P2 [12,24], and SV40 [28,39] DNA, the origin was found also on a relatively G + C rich segment.

When partial denaturation is carried out in the presence of HCHO, the degree of denaturation increases with time, and we do not, therefore, have an equilibrium state. Unfortunately, therefore, one cannot give an accurate figure for the G + C content at the origin of replication in the above examples.

IV. DISCUSSION

A. Degree of Denaturation

Successful use of denaturation mapping will depend primarily on the presence of gross base content fluctuation along the molecule. The method will be less effective for molecules that have small fluctuations in base content. Usually, the width of the equilibrium absorbancy melting profile is a good indication for the efficacy of the denaturation mapping technique [12,13,46]; if the melting curve is broad then denaturation mapping experiments will be worthwhile.

When denaturation mapping is used in studies of replication, it is necessary to choose the degree of denaturation with some care. Ideally, the molecules should be denatured to an extent that yields a characteristic denaturation pattern; this condition can only be found by trial and error. Consideration should also be given to the problems that can be encountered at high degrees of denaturation. First, as the number of denatured sites increases, the molecules become more difficult and tedious to measure. At very high degrees of denaturation the sites begin to merge together [Fig. 6(c) and (d)] and provide a less convincing frame of reference. Second, replicating molecules often have single-stranded regions associated with branch points [47,48]; in the presence of a high density of denatured sites, these two types of single strands (those originating from the replication process and those resulting from partial denaturation), can lead to erroneous interpretation. In general, therefore, it is advisable to keep the degree of denaturation as low as possible, consistent with obtaining data that yield a characteristic denaturation pattern.

In the case of 186 DNA, a characteristic denaturation pattern is observed even at 1% denaturation [Fig. 6(a)]; such a pattern would serve as an adequate frame of reference for linear molecules. At 15% denaturation [Fig. 6(b)] the pattern is sufficient to serve as a frame of reference for circular molecules. The position of the mature ends in λ and P2 circles can be convincingly determined at degrees of denaturation of 14 and 9%, respectively.

B. Position of Origin on the Genetic Map

Once the origin and direction of replication has been determined, it is of some interest to locate the position of the origin on the genetic map. Therefore, it is necessary to orient the denaturation map with respect to the genetic map. A convenient way to do this is to map the positions of a number of deletion (or insertion) mutations on the DNA molecule by heteroduplex analysis [36,37]. Denaturation mapping on the heteroduplexes will then allow a comparison of known genetic markers with the corresponding physical map [25,33].

C. Comparative

A number of studies now show that partial denaturation by either heating in the presence of HCHO, high pH in the presence of HCHO, formamide, gene 32 protein, or DNA unwinding protein from *E. coli* lead to essentially similar patterns. There are, however, obvious differences in the fine structure obtained by these methods. It appears to us that denaturation by high pH (in the presence of HCHO) yields the greatest meaningful fine structure in denaturation mapping. We believe, on the basis of preliminary experiments, that the presence of HCHO results in a complex interplay between denaturation and HCHO-mediated DNA-DNA cross-linking, and that the latter reaction plays an important part in minimizing elongation of denatured sites. If HCHO is omitted, then denaturation from ends is observed [49]. Similarly, if HCHO is present but denaturation is carried out at 4°C, then again "end effects" are observed and internal denatured sites tend to enlarge as denaturation proceeds (unpublished observations). Finally, if the DNA is highly cross-linked prior to denaturation (by preincubation with HCHO), then the denatured sites become shorter and get smaller as the preincubation time is extended (unpublished observations).

Staining of DNA has been used to improve the contrast of single strands [18,33] and a combination of staining and shadowing has also been reported (Section II,A,4 and Ref. 37). In our hands, the visualization of single-stranded DNA was not appreciably improved when carbon-coated mica discs containing the DNA were stained rather than shadowed. We should note, however, that the success of either shadowing or staining is apparently quite variable and that both methods are capable of yielding excellent results. Gene 32 protein also greatly enhances the visibility of single-stranded regions in double-stranded molecules [6,28].

ACKNOWLEDGMENTS

This work was supported by grants from the United States Public Health Service, National Institutes of Health, the American Cancer Society, and the Graduate School, University of Wisconsin.

REFERENCES

1. J. Marmur and P. Doty, J. Mol. Biol., 5, 109 (1962).

2. D. M. Crothers, J. Mol. Biol., 9, 712 (1964).

3. J. Marmur and P. Ts'o, Biochim. Biophys. Acta, 51, 32 (1961).

4. S. Lewin, Arch. Biochem. Biophys., 113, 584 (1966).

5. N. Sigal, H. Delius, T. Kornberg, M. L. Gefter, and B. Alberts, Proc. Natl. Acad. Sci. U.S., 69, 3537 (1972).

6. H. Delius, N. J. Mantell, and B. Alberts, J. Mol. Biol., 67, 341 (1972).

7. R. B. Inman, J. Mol. Biol., 18, 464 (1966).

8. R. B. Inman, J. Mol. Biol., 28, 103 (1967).

9. R. B. Inman, in Methods of Enzymology (L. Grossman and K. Moldave, eds.), Vol. 29E, Academic, New York, 1974, page 451-458.

10. R. B. Inman and M. Schnös, in Principles and Techniques of Electron Microscopy (M. A. Hayat, ed.), Vol. 4, Van Nostrand Reinhold Co., New York, 1973, in press.

11. A. Jaffe and N. Henry, Biochim. Biophys. Acta, 190, 541 (1969).

12. R. B. Inman and G. Bertani, J. Mol. Biol., 44, 533 (1969).

13. B. Gómez and D. Lang, J. Mol. Biol., 70, 239 (1972).

14. F. Mayer, W. Lotz, and D. Lang, J. Virol., (1973), 11, 946.

15. E. A. C. Follet and L. V. Crawford, J. Mol. Biol., 28, 461 (1967).

16. E. A. C. Follet and L. V. Crawford, J. Mol. Biol., 34, 565 (1968).

17. M. F. Bourguignon, Biochim. Biophys. Acta, 166, 242 (1968).

18. W. Doerfler and A. K. Kleinschmidt, J. Mol. Biol., 50, 579 (1970).

19. W. Doerfler, H. Hellmann, and A. K. Kleinschmidt, Virology, 47, 507 (1972).

20. K. Yoshiike, A. Fununo, and K. Suzuki, J. Mol. Biol., 70, 415 (1972).

21. D. R. Wolstenholme, R. G. Kirschner, and N. J. Gross, J. Cell Biol., 53, 393 (1972).

22. R. B. Inman and M. Schnös, J. Mol. Biol., 49, 93 (1970).

23. M. Schnös and R. B. Inman, J. Mol. Biol., 51, 61 (1970).

24. M. Schnös and R. B. Inman, J. Mol. Biol., 55, 21 (1971).

25. D. K. Chattoraj and R. B. Inman, J. Mol. Biol., 66, 423 (1972).

26. D. K. Chattoraj, M. Schnös, and R. B. Inman, Virology (1973), 55, 439.

27. D. K. Chattoraj and R. B. Inman, Proc. Natl. Acad. Sci. U.S., 70, 1768 (1973).

28. C. Mulder and H. Delius, Proc. Natl. Acad. Sci. U.S., 69, 3215 (1972).

29. H. B. Younghusband and A. J. D. Bellet, J. Virol., 10, 855 (1972).

30. P. C. Wensink and D. D. Brown, J. Mol. Biol., 60, 235 (1971).

31. D. D. Brown, P. C. Wensink, and E. Jordan, J. Mol. Biol., 63, 57 (1972).

32. D. D. Brown, P. C. Wensink, and E. Jordan, Proc. Natl. Acad. Sci. U.S., 68, 3175 (1971).

33. J. Wolfson, D. Dressler, and M. Magazin, Proc. Natl. Acad. Sci. U.S., 69, 499 (1972).

34. R. W. Davis and R. W. Hyman, J. Mol. Biol., 62, 287 (1971).

35. A. K. Kleinschmidt, D. Lang, D. Jacherts, and R. K. Zahn, Biochim. Biophys. Acta, 61, 857 (1962).

36. B. C. Westmoreland, W. Szybalski, and H. Ris, Science, 163, 1343 (1969).

37. R. W. Davis, M. Simon, and N. Davidson, in Methods in Enzymology (L. Grossman and K. Moldave, eds.), Vol. 21, Academic, New York, 1971, pp. 413-428.

38. D. Kaiser, in The Bacteriophage Lambda (A. D. Hershey, ed.). Cold Spring Harbor Laboratory, New York, 1971, pp. 195-210.

39. G. C. Fareed, C. F. Garon, and N. P. Salzman, J. Virol., 10, 484 (1972).

40. D. Dressler, J. Wolfson, and M. Magazin, Proc. Natl. Acad. Sci. U.S., 69, 998 (1972).

41. R. G. Wake, A. D. Kaiser, and R. B. Inman, J. Mol. Biol., 64, 519 (1972).

42. A. Skalka, M. Poonian, and P. Bartl, J. Mol. Biol., 64, 541 (1972).

43. H. Kasamatsu, D. L. Robberson, and J. Vinograd, Proc. Natl. Acad. Sci. U.S., 68, 2252 (1971).

44. H. Kasamatsu and J. Vinograd, Nature New Biol., 241, 103 (1973).

45. R. B. Inman and M. Schnös, DNA Synthesis In Vitro (R. D. Wells and R. B. Inman, eds.), University Park Press, Baltimore, 1973, page 437-449.

46. J. Geisselsoder and M. Mandel, Biophys. J., 12, 173a (1972).

47. R. B. Inman and M. Schnös, J. Mol. Biol., 56, 319 (1971).

48. J. Wolfson and D. Dressler, Proc. Natl. Acad. Sci. U.S., 69, 2682 (1972).

49. M. Fuke, A. Wada, and J. Tomizawa, J. Mol. Biol., 51, 255 (1970).

50. N. Davidson and W. Szybalski, in The Bacteriophage Lambda (A. D. Hershey, ed.), Cold Spring Harbor Laboratory, New York, 1971, pp. 45-82.

51. D. Perlman, T. Twose, M. Holland and R. H. Rownd (in preparation).

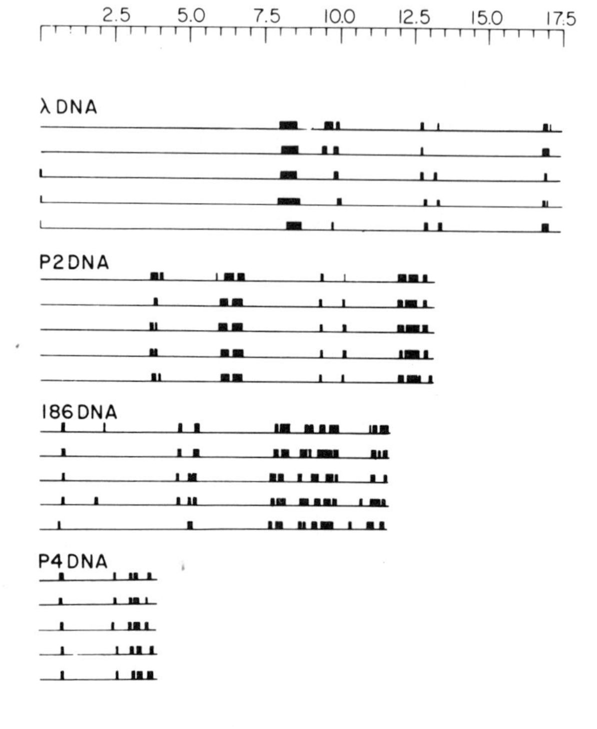

FIG. 1. Examples of phage DNA denaturation maps. Each horizontal line represents a DNA molecule and the black rectangles show the size and position of denatured sites. The maps have been normalized to the average length of the particular native DNA. The maps have been arranged so that the right halves have most denatured sites. Denaturation conditions were pH 11.28 for 10 min, pH 11.11 for 10 min, pH 11.17 for 20 min, and pH 11.09 for 10 min for λ, P2, 186, and P4 phage DNA, respectively. [Reprinted in part from Ref. 9 by courtesy of the Academic Press, New York.]

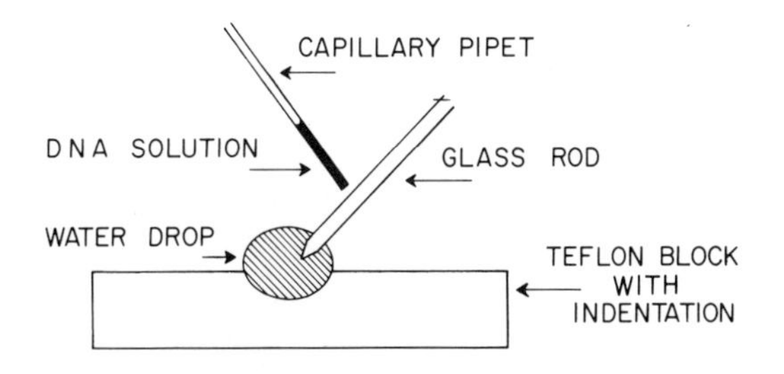

FIG. 2. Diagram showing how a DNA-containing film is formed on a water drop. [reprinted from Ref. 22 by courtesy of the Academic Press, London.]

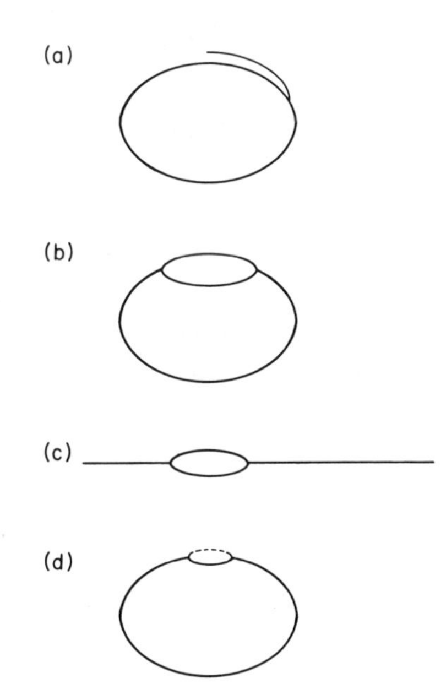

FIG. 3. Examples of different forms of intracellular phage DNA at intermediate stages of replication. (a) A single-branched circle, (b) a double-branched circle, (c) a double-branched rod, and (d) a circle with a single-stranded D-loop (the dotted line shows the single strand).

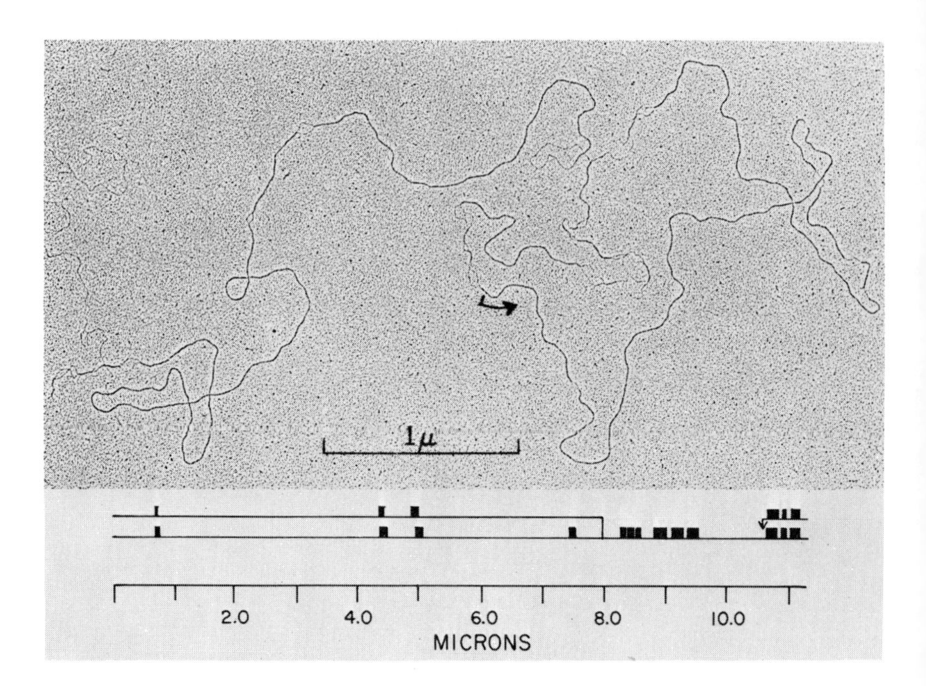

FIG. 4. Electron micrograph of a partially denatured (pH 11.21 for 10 min at 23°C) single-branched phage 186 DNA molecule. The denaturation map of the molecule is shown below the micrograph--both the circle and the branch are artificially broken, for display purposes, at a point corresponding to the mature ends of linear 186 DNA (marked by the tail of the arrow in the micrograph). From this point the denaturation map is measured in the direction of the arrow. The position of the branch point is shown by a short vertical line on the denaturation map while the vertical arrow marks the free end of the branch (the origin of replication). Single-stranded regions other than denatured loops have not been shown on the denaturation map. [Reprinted from Ref. 27 by courtesy of the National Academy of Sciences, Washington, D.C.]

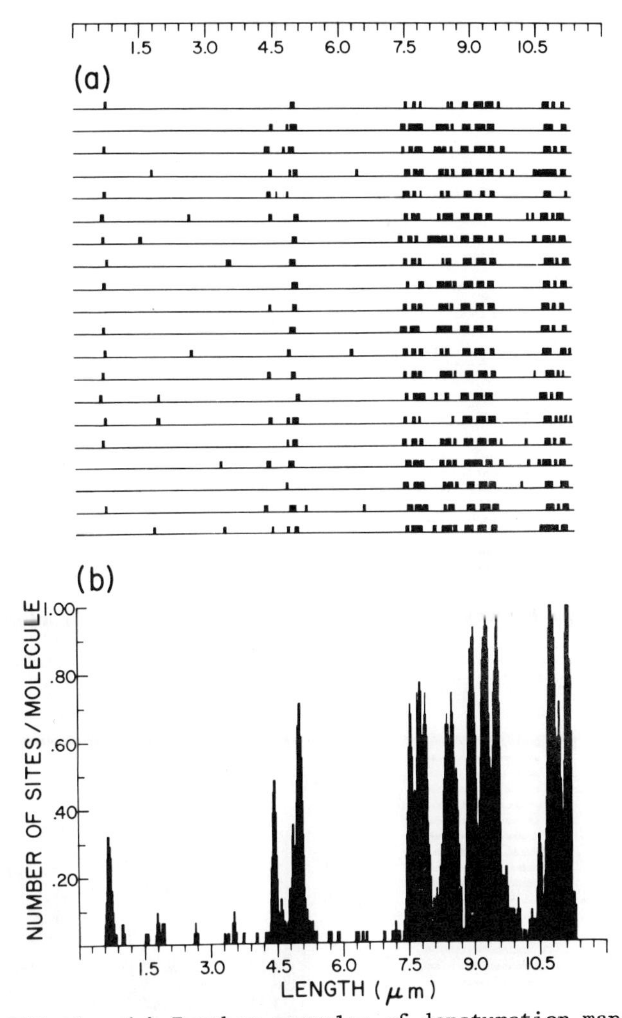

FIG. 5. (a) Further examples of denaturation maps of branched circular 186 DNA as shown in Fig. 4 (average degree of denaturation 14%). The branches have been omitted for clarity. (b) The weight-average histogram (defined in Ref. 8) of the complete set of molecules represented in (a). Thirty-nine molecules were measured. (a) and (b) are to be compared with the maps and the histogram average for the mature molecules at a similar degree of denaturation [Fig. 6(b) and (f)]. [Reprinted in part from Ref. 27 by courtesy of the National Academy of Sciences, Washington, D.C.]

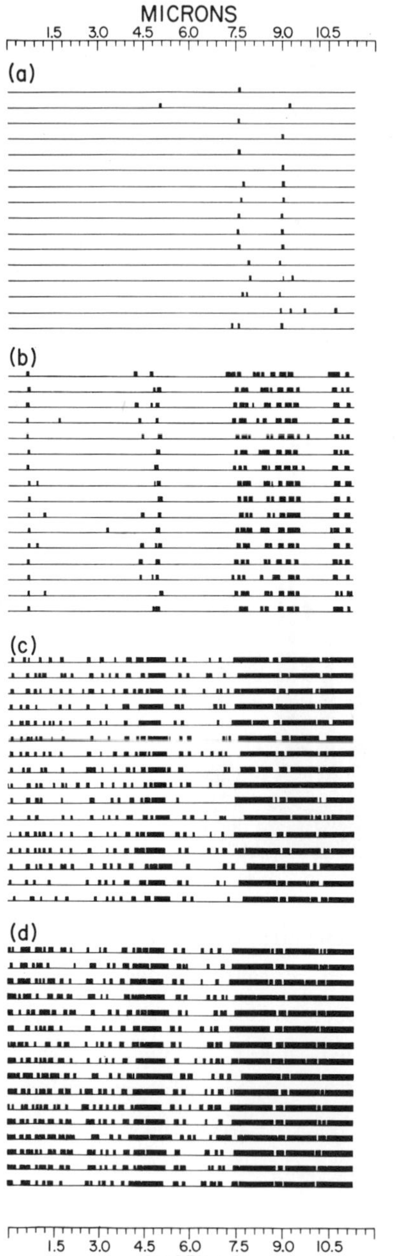
MICRONS
1.5
3.0
4.5
6.0
7.5
9.0
10.5
(a)
(b)
(c)
(d)

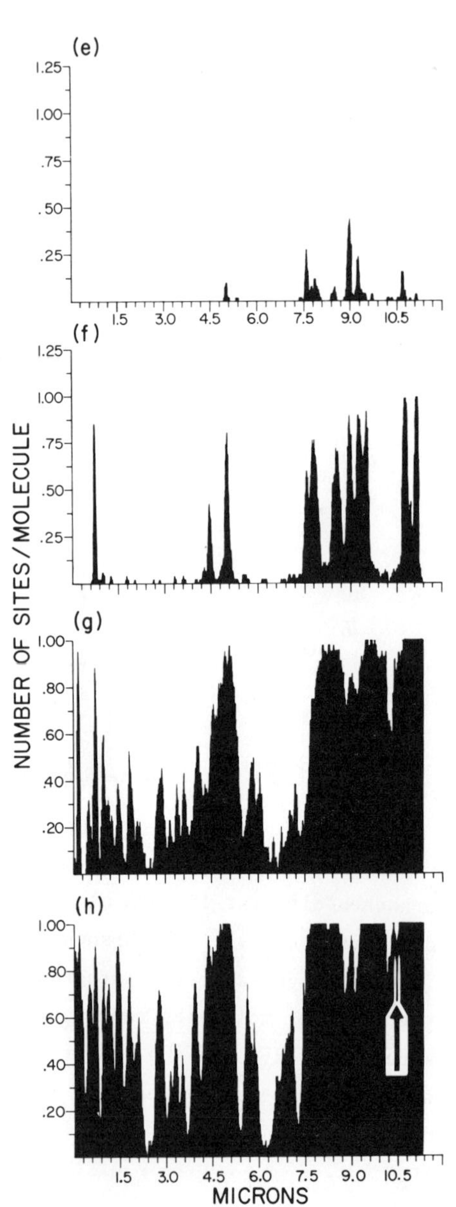
(e)
(f)
(g)
(h)
NUMBER OF SITES/MOLECULE
1.25
1.00
.75
.50
.25
.80
.60
.40
.20
1.5
3.0
4.5
6.0
7.5
9.0
10.5
MICRONS

FIG. 6. Denaturation maps and histogram averages of mature 186 phage DNA. Figure 6a-d show representative maps obtained by alkali treatment at 23°C under the following conditions: (a) pH 10.95 for 10 min then adjusted to 50% formamide. Average degree of denaturation was 1%. (b) pH 11.17 for 20 min then adjusted to 50% formamide. Average degree of denaturation was 15%. (c) pH 11.2 for 40 min then adjusted to 50% formamide. Average degree of denaturation was 45%. (d) pH 11.2 for 10 min then adjusted to 70% rather than 50% formamide. Average degree of denaturation was 58%. (e)-(h) show weight average histograms of the complete sets of molecules represented in (a)-(d) above. The number of molecules in each set was 50, 50, 40, and 43 for (e), (f), (g), and (h), respectively. All maps have been normalized to 11.3 μm, our best estimate of the length of undenatured 186 DNA. The arrow shown in (h) represents the position of the origin of replication. [Reprinted from Ref. 26 by courtesy of the Academic Press, London.]

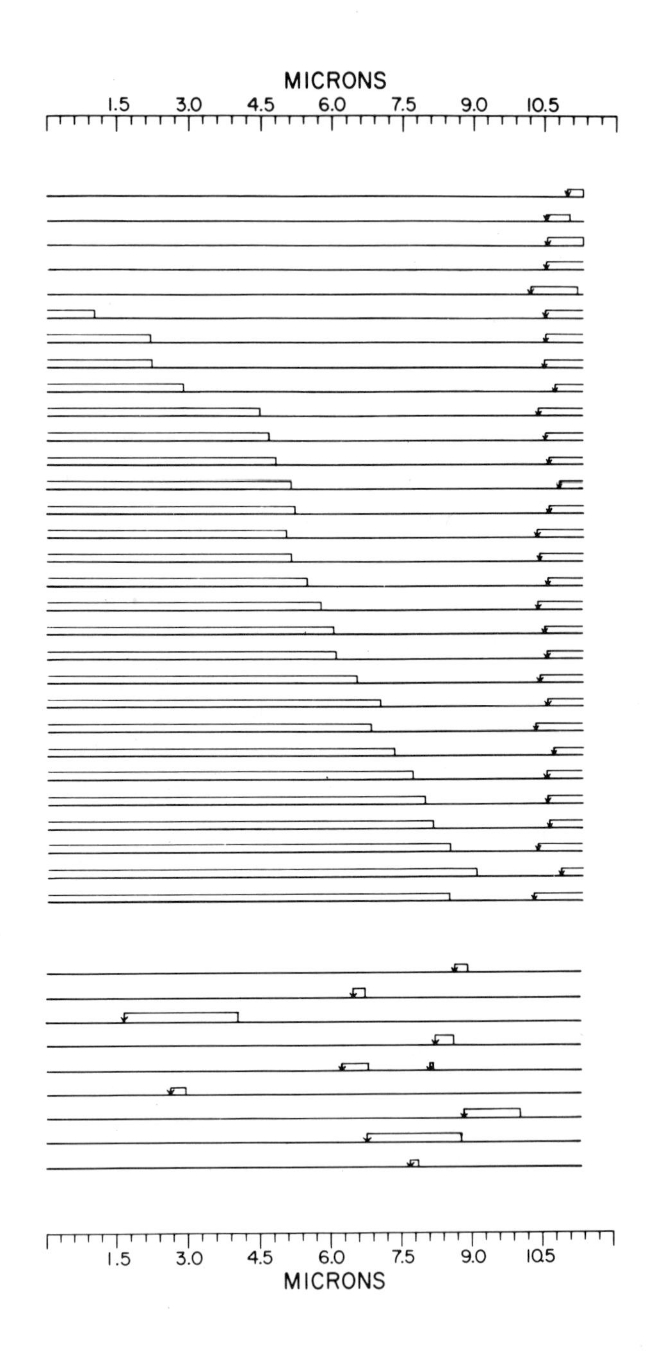
MICRONS
1.5
3.0
4.5
6.0
7.5
9.0
10.5
1.5
3.0
4.5
6.0
7.5
9.0
10.5
MICRONS

FIG. 7. Replication maps of branched circular 186 DNA molecules arranged according to increasing degree of replication (except for last nine molecules). The maps have been obtained as indicated in Fig. 4 except that the denatured sites have been omitted for simplicity. The branches in the third to fifth molecule from the top are entirely single stranded. The free end of these branches thus cannot be located by denaturation mapping. However, when these branches are oriented in the same direction as the remaining molecules, they are found to yield an unique origin of replication. The same was not true for the last nine molecules, four of which were entirely single stranded. [Reprinted from Ref. 27 by courtesy of The National Academy of Sciences, Washington, D.C.]

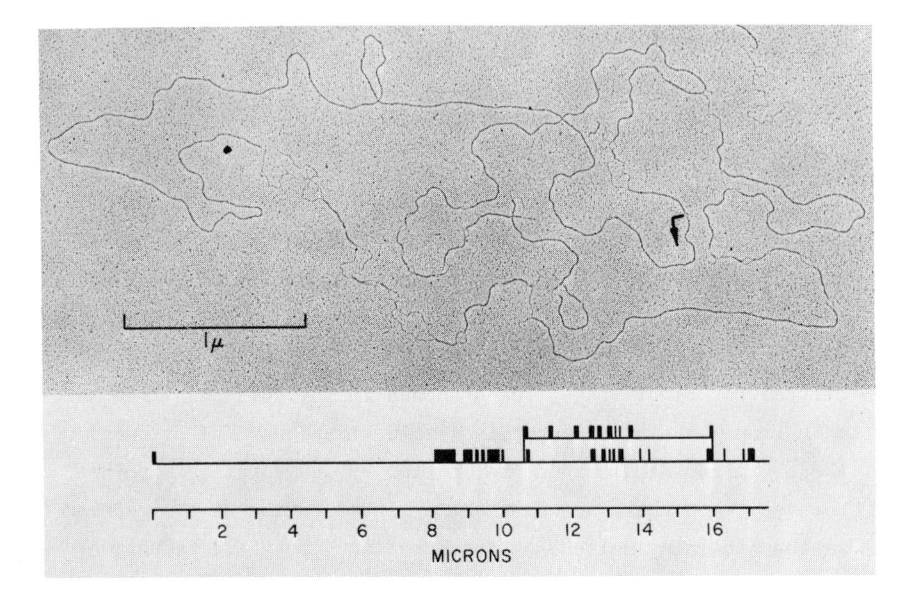

FIG. 8. Electron micrograph of a partially denatured replicating λ DNA molecule (pH 11.05 for 10 min). As in Fig. 4, the denaturation map is measured in the direction of the arrow starting from a point corresponding to the mature ends of λ DNA (marked by the tail of the arrow). The ends of the double denaturation map (representing the daughter segments) indicate the position of the two replicating growing points, the mid-point of which is used to find the origin of replication [Fig. 9(a)]. [Reprinted from Ref. 23 by courtesy of the Academic Press, London.]

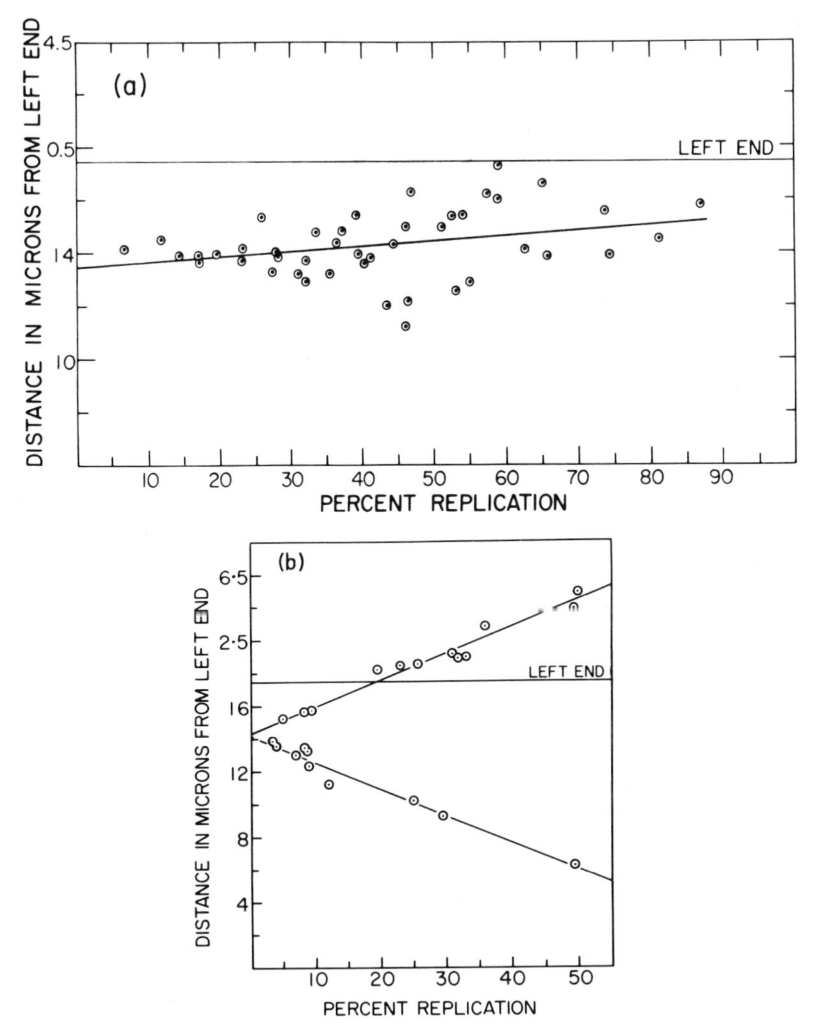

FIG. 9. (a) Distribution of the position of midpoints of daughter segments in λ DNA molecules (which replicate bi-directionally) plotted against percentage replication (calculated from the length of the daughter segments). The mean line is drawn by a best-fit calculation. (b) Distribution of the position of the growing point in λ DNA molecules, which replicate uni-directionally. The upper best-fit line is drawn through the points that proceed to the right and the lower line for the points that proceed to the left. [Reprinted in part from Ref. 23 by courtesy of the Academic Press, London.]

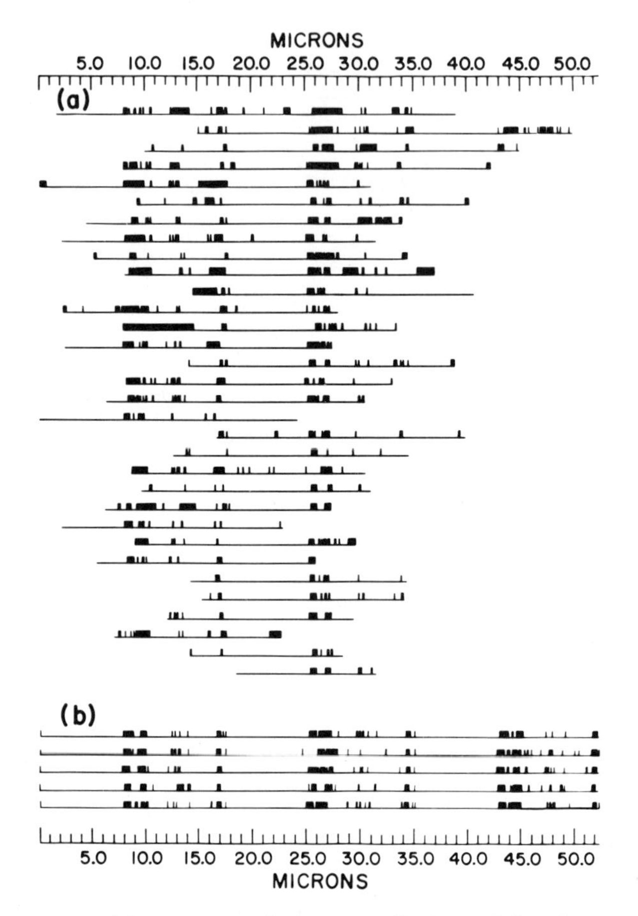

FIG. 10. (a) Denaturation maps of phage λ head mutant DNA (λE III). Maps have been aligned to give the best visual fit of the denatured sites with the λ control [shown in (b)]. (b) Denaturation maps of computed head-to-tail trimers of λcI857 DNA. Each horizontal line represents the expected map resulting from the head-to-tail union of three real monomeric λ denaturation maps. The position of mature ends are located at 0.0, 17.5, 35.0, and 52.5 μm. [Reprinted from Ref. 41 by courtesy of the Academic Press, London.]

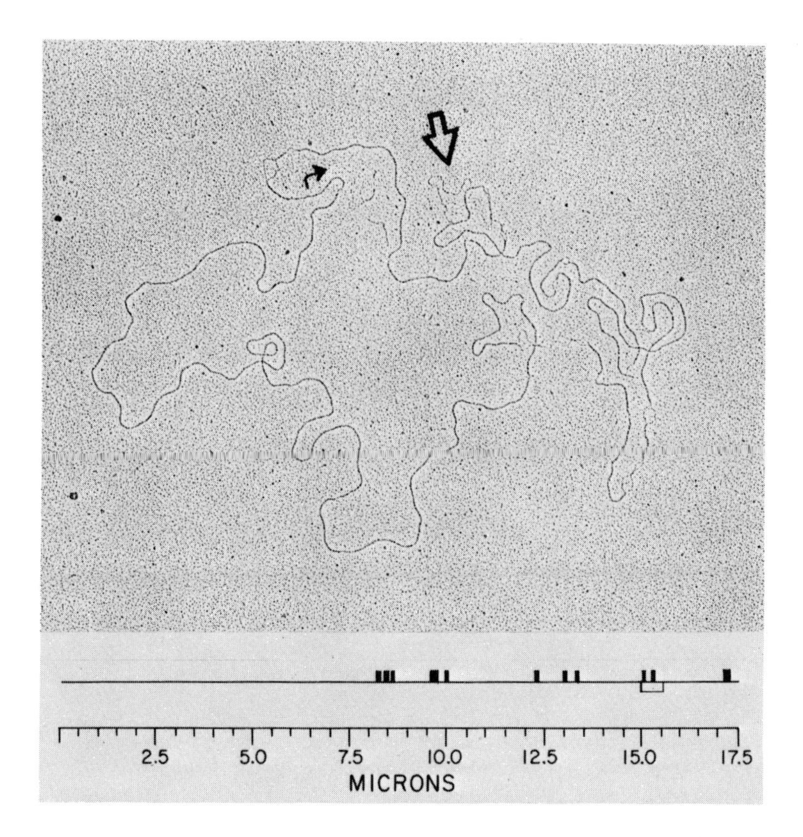

FIG. 11. Electron micrograph of a partially denatured λ DNA circle containing a D-loop (open arrow). Its position can be deduced from the denaturation map that was measured from a point corresponding to the mature ends of λ DNA (closed arrow). The size and position of the D-loop is shown by the open rectangle below the denaturation map. [Reprinted from Ref. 10 by courtesy of Van Nostrand Reinhold Co., New York.]

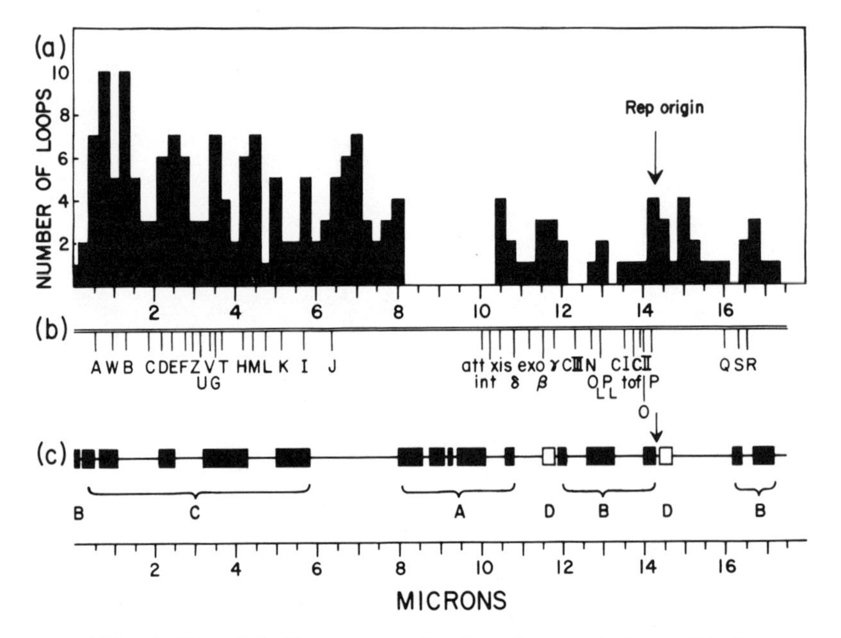

FIG. 2-12. (a) Histogram showing the position of D-loops with respect to mature ends of phage λ DNA. Denaturation mapping was used as a frame of reference to determine the position of the mature ends. (b) The physical position of various λ genes [50]. (c) Grand average denaturation map of λ. The A + T content decreases in segments A, B, and C, respectively. The positions marked D represent the G + C richest regions in the right half of the molecule. The origin of replication is shown by the arrow. [Reprinted from Ref. 45 by courtesy of University Park Press, Baltimore.]

Chapter 3

RESTRICTION ENDONUCLEASE FROM *HEMOPHILUS INFLUENZAE* RD

Hamilton O. Smith

Department of Microbiology
The Johns Hopkins University
School of Medicine
Baltimore, Maryland

I. INTRODUCTION

A growing number of molecular biologists are turning to restriction enzymes as tools for analysis of DNA molecules and chromosomal DNA. Restriction enzymes are site-specific endonucleases that can be isolated from various strains of bacteria. They are capable of recognizing particular base sequences within native DNA molecules and producing double-strand cleavage. Thus viral genomes or other DNA molecules may be cleaved in a highly specific and reproducible manner into a number of double-stranded fragments. These are useful for localization of genetic functions, DNA base sequencing, and a number of other purposes.

This article describes the purification of restriction enzyme from *H. influenzae* Rd. The procedure is similar to that reported by Smith and Wilcox [1] but incorporates modifications which facilitate assay and large scale purification.

II. ASSAYS

Restriction endonucleases, because of their high specificity, cleave DNA to only a limited extent. Little if any acid-soluble material is released. Therefore most of the assay procedures employ sucrose gradient analysis [2] or other techniques capable of detecting only a few double-strand breaks in the substrate DNA. These assays are sensitive, but are often time consuming and difficult to quantitate. The two relatively simple methods described below have been extensively used by the author. The first is a rapid qualitative method that is useful for locating activity among a series of fractions. The second may be used to quantitate the amount of activity.

A. Clot Assay

This qualitative assay for restriction enzyme is based on the principle that high-molecular-weight DNA forms a compact clot-like precipitate in the presence of small amounts of bovine serum albumin (BSA) and 5% trichloroacetic acid (TCA), while low-molecular-weight DNA formed by cleavage with restriction enzyme yields a finely divided, "milky" precipitate.

1. Reagents

TSM buffer: 10 mM Tris-HCl, pH 7.5, 50 mM NaCl, 10 mM $MgCl_2$.

TCA solution: 10% (w/v).

BSA solution: 0.3% (w/v).

Salmon sperm DNA solution: Highly polymerized salmon sperm DNA (Sigma) is dissolved at 2 mg/ml in 10 mM Tris-HCl, pH 8, and extracted with an equal volume of phenol equilibrated with 100 mM Tris-HCl, pH 8. The aqueous phase is brought to 0.1 M NaCl by addition of 5 M NaCl and precipitated with 2 vol of 95% ethanol. The precipitate is wound onto a glass rod, rinsed thoroughly in three separate tubes, each containing 10 ml of 95% ethanol, and redissolved in the original volume of 10 mM Tris-HCl, pH 7.5. The concentration of DNA will be about 1.5 mg/ml by A_{260} measurement. (An A_{260} of 1.0 = 50 μg of DNA/ml.)

2. Procedure

The reaction mixture contains:

TSM buffer	0.1 ml
Salmon sperm DNA	0.02 ml
Enzyme or extract	1-20 μl

Incubate at 37°C for 20 min. Add 1 drop (0.05 ml) of BSA solution and 0.2 ml of 10% TCA solution. After mixing, the assay result may be read immediately. A milky, fine precipitate is positive for restriction activity, and a clotted DNA precipitate is negative. Less than 0.05 units of enzyme, as defined below, is readily detectable. The amount of enzyme can be roughly quantitated by running a series of assays in which either the time of incubation or the amount of enzyme is varied. In cases where there is doubt as to whether restriction endonuclease is being measured, a control reaction should be carried out using H. influenzae Rd DNA. Restriction enzyme is inactive on the fully modified host DNA.

B. DNA Viscometry Assay

The intrinsic viscosity of native DNA increases as the DNA chain length increases [3]. Endonucleases rapidly reduce the average length and, consequently, the viscosity of DNA, while exonucleases remove nucleotides stepwise from the ends, resulting in very little initial effect on the viscosity. Therefore, measurement of decrease in DNA viscosity provides a sensitive assay for restriction endonuclease activity. It is, of course, essential that the extract or preparation to be assayed contain no significant nonspecific endonucleases. This is the case in H. influenzae.

1. Reagents

TSM buffer

Salmonella phage P22 DNA solution: Phage P22 is purified from L broth lysates [4] by differential centrifugation and by banding in CsCl step gradients [5]. The DNA is extracted with

1 vol of cold, redistilled phenol and then precipitated with 2 vol of cold ethanol and redissolved in 1 vol of NaCl-Tris buffer (0.05 M NaCl, 0.01 M Tris-HCl, pH 7.4). The precipitation is repeated, and the DNA is finally redissolved in the above buffer at 1.30 mg/ml.

Salmon sperm and calf thymus DNA are suitable substitutes if highly polymerized. They should be extracted with phenol before use to inactivate and remove contaminating nucleases found in some commercial preparations. These DNAs generally are supplied at molecular weights of 10×10^6 or less, whereas the molecular weight of phage P22 DNA is 26×10^6. Their viscosity per equivalent DNA concentration is therefore less and the assay sensitivity is proportionately lower. Bacterial or other high-molecular-weight viral DNAs, comparable to phage P22 DNA, are also suitable.

2. Procedure

DNA viscosity measurements may be carried out using a simple Ostwald viscometer of the type supplied by A. H. Thomas Co., Philadelphia (Cat. No. 7162-N10) having a flow time for water of approximately 60 sec. The viscometer is operated in a thermostated glass chamber at 30°C. One may conveniently use a 4-liter beaker filled with water and heated by a Brownwill Model 20 constant temperature circulator.

The viscometer is filled with 3.5 ml of phage P22 DNA solution, 40 μg/ml in TSM buffer. The capillary side is fitted with a short length of rubber tubing so that the solution may be drawn by oral suction into the measuring (upper) bulb. A measurement consists of determining the time for the solution meniscus to pass from the first engraved line (just above the measuring bulb) to the second engraved line (just below the bulb). Several flow-time measurements are taken after the DNA solution has reached thermal equilibrium and these are generally reproducible to within 0.1 sec. The P22 DNA produces an approximately 15% increment in the flow time compared with pure solvent. Five to 50 μl of extract or purified enzyme is introduced into the reservoir bulb of the viscometer and rapidly mixed by blowing air retrograde into the bulb. Flow-time measurements are taken as rapidly as is practical. The DNA viscosity is expressed as specific viscosity, $\eta_{sp} = (t/t_D) - 1$, where t_D represents the flow time at the end of the experiment, 5 min after addition of 0.05 ml of pancreatic DNase, 1 mg/ml in 0.01 N HCl (this corresponds to the flow time for pure solvent within the accuracy of the measurements). Specific viscosity measurements are plotted against time on semilogarithm paper as fractional values of the zero-time value. One unit of enzyme activity is defined as that amount that produces a decrease in the DNA specific viscosity of 25% in 1 min under the conditions described above. It should be pointed out that the viscometric assay is valid even on crude extracts since no activity against *H. influenzae* Rd DNA is found in crude extracts. It should be noted that three units as defined here are approximately equal to one unit as previously defined [1] because 50 mM NaCl is included in the buffer.

III. PURIFICATION OF RESTRICTION ENDONUCLEASE

A. Growth of Cells

1. Medium

NAD (β-nicotinamide adenine nucleotide, Sigma Chem. Co., Grade III): 10 mg/ml. Sterilize by filtration through a 0.45 μm Millipore filter. Store at 4°C.

Hemin (Eastman, Cat. No. 2203), 0.1% in 4% 2, 2', 2"-nitriloethanol (triethanolamine): Dissolve and sterilize by heating in a waterbath at 70°C for 30 min.

Brain-heart infusion (Difco) broth, 3.7%: Autoclave 15 min at 121°C. We and others [6] have observed that prolonged autoclaving diminishes the ability of the medium to support *H. influenzae* growth.

Supplemented brain-heart infusion (sBHI) broth contains 10 ml of hemin solution and 0.2 ml of NAD solution per liter.

2. Storage of Cultures

H. influenzae is autolytic and consequently does not survive well in stationary cultures. Agar slants or stab cultures survive for less than a week. Therefore, it is necessary to maintain frozen stocks. Cells are grown in sBHI broth on a shaker bath at 37°C to an optical density at 650 nm of 0.65 (approximately 10^9 cells/ml). Glycerol (autoclaved) is added to 17% and mixed thoroughly. Aliquots of cell suspension are distributed into vials and stored frozen at -60° or below. Viability is maintained for many months.

3. Procedure

Small quantities (up to 30 g wet weight) of cells may conveniently be grown in shaker flasks. Larger quantities are obtained by growth under controlled conditions in a fermentor. (Note: The Rd strain is a nonencapsulated, nonpathogenic strain that is routinely handled in our laboratory with no greater precautions than are exercised for *E. coli*.) For maximum yield of cells in the fermentor, the following procedure has been adopted although numerous variations are possible depending on available equipment.

A New Brunswick Model FS314 three-chamber fermentor is used. Eleven liters of distilled water and 400 g of brain-heart infusion powder are added to each 15-liter jar and dissolved thoroughly prior to autoclaving. The fermentor chambers are assembled and autoclaved for 30 min using an American Sterilizer Co. Model 57CR autoclave. (Note: Some spores survive this amount of autoclaving. Use the medium immediately after autoclaving.) The complete cycle takes about 90 min. Remove the hot chambers, place them in the fermentor waterbath containing cool water, and begin slow stirring action. They should cool to 37°C in about 1 hr. Set the waterbath thermostat at 37°C. Add 110 ml hemin solution and 2.5 ml NAD solution to each chamber. Inoculate each jar with 500 ml of late log phase cells which were obtained by inoculating each of three 2-liter shaker flasks containing 500 ml of sBHI with 50 ml of frozen log phase cells about 3 hr previously. Begin stirring at 250-300 rpm and regulate the aeration to 10-12 liter min. Foaming is controlled by intermittent addition of Dow Antifoam B emulsion (diluted with nine parts of water). The cells usually reach an optical density at 650 nm from 1.7 to 2.0 within 5 hr after inoculation (Doubling time, 30 min). Less than 250 ml of diluted antifoam is usually required per jar. The cells are harvested using a Sharples centrifuge. Yield from the 36 liters of culture is approximately 180-200 g of cell paste. The level of contamination usually does not exceed 10^{-6}. The cells are stored at -60°C until used.

B. Preparation of Cell Extract

Frozen cells (330 g) are thawed and suspended in 1100 ml of 25 mM Tris·HCl, pH 8, 5 mM 2-mercaptoethanol. Aliquots of 400 ml are placed in a 600-ml stainless-steel beaker cooled in an ice-salt-water bath. The cells are disrupted by continuous sonication at the full intensity (8 A) of a Branson sonicator. Temperature is monitored by a metal thermometer and should remain below 8°C during the entire procedure. Turbidity measurements are followed at 10-min intervals. Disruption is sufficiently complete when the turbidity falls to 15-20% of starting values (usually after 30-60 min of sonication). Other methods of cell disruption have not been explored. Cell debris is removed by centrifugation in a Sorvall GSA rotor at 12,000 rpm (22,000 *g*) for 60 min. Approximately 1350 ml of "turbid" supernatant is obtained.

C. Restriction Enzyme Purification Procedure

All operations are carried out at 0°-4°C.

1. Streptomycin Precipitation

The cell extract (1350 ml, 290 A_{260} units/ml, A_{230}/A_{260} = 1.3) is stirred on a magnetic stirrer and 220 ml of 10% streptomycin sulfate solution (freshly prepared) is added dropwise from a separatory funnel over a 15-30 min period (1 ml per 1800 A_{260} units). The suspension is allowed to stand overnight. The precipitate is removed by centrifugation at 22,000 *g* for 30 min. The clear amber supernatant (1325 ml), A_{230}/A_{260} = 2.1 to 2.5, is recovered. An absorbance ratio less than 2.0 indicates that nucleic acids have not been well removed.

2. Ammonium Sulfate Precipitation

Dilute the streptomycin supernatant to 3 liters with 25 mM Tris·HCl, pH 8, 5 mM 2-mercaptoethanol, 0.3 M NaCl. Add 939 g of ammonium sulfate (50% of saturation) over a 30-min period. After an additional 30 min, remove the precipitate by centrifugation. To the supernatant, add 477 g of ammonium sulfate (70% of saturation) and collect the precipitate. Redissolve the precipitate in 40 ml (or in as small a volume as possible) of 0.01 M sodium phosphate buffer, pH 7.4.

3. Sephadex G-25 Desalting

The 50-70% ammonium sulfate precipitate (77 ml) is layered carefully onto the surface of a G-25 Sephadex column (4 x 37 cm, 460-ml bed volume) equilibrated with 0.01 M sodium phosphate buffer, pH 7.4. The sample volume to column bed volume ratio is best kept at about 1:5. The column is eluted rapidly and followed visually. The main excluded protein band is amber colored and is followed by a clear zone and a trailing light yellow band marking the position of low-molecular-weight compounds and salts. The main protein band is collected in a volume of about 120 ml.

4. Phosphocellulose Chromatography

Whatman P11 phosphocellulose (100 g) is suspended in 2 liters of 0.5 N NaOH, stirred for 5 min, and allowed to stand at room temperature for 30 min. The supernatant is decanted, and the phosphocellulose is collected in a large Buchner funnel and is washed extensively with water, 0.5 N HCl, water, and finally with 0.1 M sodium phosphate, buffer, pH 7.4. The cellulose suspension is retitrated if necessary to pH 7.4 with 0.5 M Na_2HPO_4 solution. It is stored at 4°C in the buffer.

A 2.5 cm x 12 cm (60 ml bed volume) column is poured and

equilibrated to pH 7.4 overnight with 0.01 M sodium phosphate buffer, pH 7.4 at a flow rate of 75 ml/hr. The desalted ammonium sulfate 50-70% precipitate fraction is loaded at 65 ml/hr. The column is eluted stepwise at 65 ml/hr with 120 ml of equilibration buffer, 120 ml of 0.1 M NaCl in buffer, 240 ml of 0.2 M NaCl in buffer, and 270 ml of 0.3 M NaCl in buffer. Fractions (6.5 ml) are collected during the final two steps. Activity, as measured by the clot assay, generally elutes in the interval of 1 to 3 column volumes of the 0.3 M NaCl step. Gradient elution (0.15 M to 0.4 M) is also satisfactory.

The individual fractions (averaging 10-20 units/ml of restriction activity) should be assayed also for exonucleolytic activity as described by Smith and Wilcox [1]. Those containing significant exonuclease activity should be stored separately. Other fractions may be pooled if desired.

5. Concentration and Storage of Enzyme

Even dilute enzyme (5 units/ml) can be stored in 50% glycerol at -20°C for over a year without loss of activity. The dilute phosphocellulose enzyme fraction is relatively stable at 4°C, and in the presence of added 0.3% bovine serum albumin (Sigma) is stable for many months.

The enzyme may be concentrated about threefold in one step using lyphogel (Gelman Instrument Co., Ann Arbor, Mich.) or concentrated by reprecipitation with ammonium sulfate or by threefold dilution to 0.1 M NaCl concentration followed by adsorption to a 1 ml phosphocellulose column and elution in a small volume (less than 1 ml) with 0.5 M NaCl. The latter procedures, however, often result in 50% or greater losses of activity. Procedures involving dialysis have resulted in 80% loss of activity.

IV. PROPERTIES

A. Specificity

The phosphocellulose fraction of H. influenzae Rd restriction enzyme (see Table 1) has proven useful for analysis of several small viral DNA genomes such as SV40 [7] and ØX-174 [8]. SV40 DNA is cleaved into 11 distinct fragments separable by polyacrylamide gel electrophoresis [9]. Lee, Danna, Smith, and Nathans [10] recently have demonstrated that the phosphocellulose fraction contains two separate restriction activities, one producing five specific breaks in SV40 DNA (endo R. Hin dII) and the other producing six breaks (endo R. Hin dIII).* Preliminary observations indicate that these activities may be separated by chromatography on DE52 DEAE-cellulose (Whatman). Phosphocellulose fraction placed in 0.01 M Tris-HCl buffer by Sephadex G-25 chromatography is loaded onto a DE52 column at 2 mg of protein per milliliter of column-bed volume. Endo R. Hin dIII appears in the 0.01 M Tris-HCl, pH 7.4 breakthrough volume [10], while endo R. Hin dII elutes at 0.08 to 0.10 M NaCl [11].

The endo R. Hin dII activity has the site specificity (5') GTPy↓PuAC (3') as determined by Kelly and Smith [12]. Phage T7 DNA, which was used for the sequence determination, is an excellent substrate for endo R. Hin dII but does not appear to be a substrate for endo R. Hin dIII.

The observations with SV40 DNA emphasize the value of assaying restriction activity by using specific DNA substrate and analyzing cleavage products by gel electrophoresis. This approach has been used recently [13] to detect and separate two H. para-

---

*Nomenclature suggested by Smith, H.O. and Nathans, D., J. Mol. Biol., 81, 419 (1973).

TABLE 1

H. influenzae Rd Restriction Enzyme Purification Chart

Fraction	Volume, ml	Total protein, mg	Activity units/ml	Activity units/mg	Activity total units	Relative recovery
I. 12,000 *g* supernatant	1350	44,500	7.1	0.22	9600	1.0
II. Streptomycin supernatant	1325	26,500	3.7	0.19	4900	0.48
III. Ammonium sulfate, 50-70% precipitate	77	6,920	35	0.39	2660	0.28
IV. Phosphocellulose	169	169	11	11	1860	0.19

influenzae restriction activities originally attributed to a single enzyme [14]. The "clot" assay and the viscometry assay reveal overall restriction activity but generally cannot distinguish the contributions to the activity arising from several different enzymes. However, by initially using a complex DNA such as that from calf thymus or salmon sperm to survey column fractions for endonucleolytic activity, one can detect enzyme that may not be active against specific viral DNAs.

B. Stability

As already indicated, the enzyme preparation is quite stable under several conditions of storage. Activity is also maintained for several hours in reaction mixtures, making possible the use of a small amount of enzyme for carrying out digestions. The enzyme is inactivated by treatment at 60°C for 10 min.

C. Ionic Requirements

Activity is optimal at 0.06 M NaCl in 7 mM $MgCl_2$; 7 mM Tris·HCl, pH 7.4. There is a threefold lower activity in the absence of added NaCl and an inhibition of activity above 0.1 M NaCl. No activity is detectable in the absence of Mg^{2+} ion, hence reactions may be terminated with chelating agents, e.g., 0.02 M ethylenediamine tetraacetic acid (EDTA).

REFERENCES

1. H. O. Smith and K. W. Wilcox, J. Mol. Biol., 51, 379 (1970).

2. M. Meselson, Procedures in Nucleic Acid Research (G. L. Cantoni and D. R. Davies, eds.), Vol. 2, Harper and Row, New York, 1971, pp. 889-895.

3. B. H. Zimm, Procedures in Nucleic Acid Research (G. L. Cantoni and D. R. Davies, eds.), Vol. 2, Harper and Row, New York, 1971, pp. 245-261.

4. M. Levine, Virology, 3, 203 (1957).

5. C. A. Thomas and J. A. Abelson, Procedures in Nucleic Acid Research (G. L. Cantoni and D. R. Davies, eds.), Vol. 1, Harper and Row, New York, pp. 553.

6. J. K. Setlow, D. C. Brown, M. E. Boling, A. Mattingly, and M. P. Gordon, J. Bacteriol., 95, 546 (1968).

7. K. J. Danna and D. Nathans, Proc. Natl. Acad. Sci. U.S., 69, 3097 (1972).

8. M. H. Edgell, C. A. Hutchison, III, and M. Sclair, J. Virology, 9, 574 (1972).

9. K. Danna and D. Nathans, Proc. Natl. Acad. Sci. U.S., 68, 2913 (1971).

10. T. Lee, K. Danna, H. Smith, and D. Nathans, private communication, 1973.

11. H. O. Smith, unpublished work, 1973.

12. T. J. Kelly, Jr., and H. O. Smith, J. Mol. Biol., 51, 393 (1970).

13. P. A. Sharp, B. Sugden, and J. Sambrook, Biochemistry, 12, 3055, (1973).

14. R. Gromkova and S. H. Goodgal, J. Bacteriol., 109, 987 (1972)

Chapter 4

THE EcoRI RESTRICTION ENDONUCLEASE

P. J. Greene, M. C. Betlach,
and Herbert W. Boyer

Department of Microbiology

Howard M. Goodman

Department of Biochemistry and Physics
University of California
San Francisco, California

I. INTRODUCTION

Type II bacterial restriction endonucleases have become increasingly useful in the analysis of small DNA genomes [1-10], while attempts to use the Type I restriction endonucleases have not been satisfactory because they appear to be less specific than originally imagined. The Type I and Type II restriction endonucleases were originally defined on the basis of size and cofactor requirements [11]. Type I restriction endonucleases (e.g., Eco<u>B</u>, EcoK, and EcoPI*) have large molecular weights and contain three different protein subunits, two of which are found in the related modification methylase [12-14]. They either require or are stimulated by Mg^{2+}, S-adenosyl-L-methionine, and adenosinetriphosphate. Type II restriction endonucleases (e.g., Hin, Hpa, EcoRI, EcoRII) only require Mg^{2+} for endonucleolytic activity, have molecular weights <100,000 and appear to be dimers or tetramers of identical protein subunits [15-17]. Endonucleolytic phosphodiester bond cleavages made by Type II enzymes occur at defined nucleotide sequences, and specific DNA fragments can be obtained [1],[5]-[7],[18]-[21]. These same sequences are methylated by the related modification methylase that renders the sequence insensitive to the endonuclease [21,22]. On the other hand, Type I endonucleases may not necessarily make endonucleolytic cleavages at defined nucleotide sequences, although interaction of Type I endonuclease with a specific unmethylated nucleotide sequence appears to be necessary as a

---

*E. coli = Eco. B, K, PI, etc., designate the strain. Hemophilus influenzae = Hin. H. Parainfluenzae = Hpa. These abbreviations are based on nomenclature for host modification and restriction systems proposed by K. D. Danna, G. Sack, and D. Nathans, personal communication.

first step [23,24]. This property of the Type I restriction endonucleases may limit their applicability in the analysis of DNA genomes. There are two properties of the EcoRI restriction endonuclease, a type II endonuclease, that are important in defining the usefulness of the endonuclease for analyzing DNA [19,20]. First of all, the EcoRI endonuclease has the highest degree of specificity of the known restriction endonucleases. The five double-strand sites cleaved in λ DNA by the EcoRI endonuclease have the following sequence:

```
              ↓
5'-----T/A G pA A T T C A/T-----3'
3'-----A/T C T T A Ap G T/A-----5'
                     ↑
```

(The arrow indicates the position of the phosphodiester bond cleavages.) It is not known yet if the endonuclease will function if a G·C base pair is at either or both of the outside positions of the sequence. If an A·T base pair is required as shown above, the EcoRI endonuclease would make, on the average, one double-strand break every 16,000 nucleotide base pairs. The minimum unique length for the sequence is six base pairs, so in either case relatively large fragments of DNA would be generated from any susceptible DNA molecule (see Table 1 for some representative data). The second property of the EcoRI endonuclease that is useful is that cohesive termini are made by the endonuclease. Any two fragments of DNA generated by the EcoRI endonuclease can reassociate through the cohesive termini and be ligated with polynucleotide ligase.

The current purification procedure for the EcoRI restriction endonuclease and some of its properties, as well as a procedure for partial purification of the related EcoRI modification methylase, are described below.

TABLE 1

EcoRI Substrate Sites on Various DNA Molecules

Source	Hits	References
SV40	1	6,7
λ	5	19,25
λ cm	4	26
Ø 80	10	26
λ plac 5	5	26
λ pbio 1	4	26
PI	14	26
P22	7	26
R6(5) plasmid	12	26
Tc6(5) (PSC101)	1	26
JA Sex	1	26
Phage G4 and G14	1	27
Polyoma	1	20
PM2	0	20
Mouse mitochondrial	2	20
Adenovirus 2	5	20
F_8(p17)	±19	20

II. BACTERIAL STRAINS

The bacterial strain (RY13) used as a source for the EcoRI restriction endonuclease and modification methylase and its construction has been described previously [28]. It is a derivative of a widely used *E. coli* 1100 (K12) line that is deficient for endonuclease I. The strain was made r_B^+ m_B^+ and a derepressed *fi*$^+$ R factor carrying the EcoRI restriction and modification genes introduced by conjugation and selection for the appropriate drug resistance genes. The original restriction and modification genes were isolated from a clinical specimen of *E. coli* carrying an *fi*$^+$ R factor. This plasmid spontaneously lost its capacity for conjugal transfer but was subsequently recombined with a derepressed *fi*$^+$ plasmid. The R factor in the RY13 strain carries drug resistance genes for sulfathiazole and streptomycin. There is an autosomal streptomycin resistant gene as well. This strain restricts unmodified λ bacteriophage (phage propagated on the F^- parental strain HB129 of RY13, which does not carry the *fi*$^+$ R factor) with an efficiency of plating of 10^{-4}. Even though we have detected less than 5% loss of the R factor drug resistance markers after as many as 15 generations, it is advisable to test single colony isolates of strain RY13 for the presence of the *Eco*RI restriction and modification properties prior to preparing large quantities of cells for purification of the *Eco*RI restriction endonuclease and methylase. The efficiency of plating of unmodified λ phage is the simplest procedure for this determination. Plate stocks of *Eco*RI modified (λ·B-RI) and unmodified (λ·B) λ phage can be prepared by standard procedures by growth in RY13 (r_B^+ m_B^+, r_{RI}^+ m_{RI}^+) and HB129 (r_B^+ m_B^+), respectively [25]. Titers for the λ·B-RI and λ·B stocks of phage are obtained by plating suitable dilutions on cultures of HB129. The titer of λ·B-RI stock should be the same on cultures of the RY13 and HB129 strains, but the λ·B stock titer is 10^{-4} on the RY13 strain compared with the HB129 strain because of the *Eco*RI restriction mechanism.

III. METHYLASE AND ENDONUCLEASE ASSAYS

The methylase is assayed in 100 μl reactions containing 100 mM Tris-HCl, pH 8.0, 10 mM EDTA, 1.1 μM [^{3}H]SAM (stored in 0.01 N H_2SO_4), 400 μg/ml BSA, and about 0.37 pmoles of λ genome (11 μg). The λ DNA solution is sonicated for 10 sec to reduce the viscosity of the solution. High-molecular-weight DNA (e.g., linear aggregates of λ DNA) significantly reduces the rate of methylation presumably because of the increased viscosity of the solution. Reactions are incubated at 37°C and stopped by placing them on ice with the addition of 1.0 ml of 7.5% cold perchloric acid with calf thymus DNA (80 μg) as carrier. After 30 min at 4°C, the acid precipitate is collected on glass fiber filters by vacuum aspiration and washed with 10 ml of 3.5% ice-cold perchloric acid, 8 ml of ice cold 2 N HCl, and finally with 5 ml of 95% ethanol. After drying the filters under a heat lamp, tritium decay is counted in 10 ml of Omnitol [4 g of Omnifluor (NEN) per liter of toluene] scintillation fluid in a Beckman LS200 B scintillation counter. Specific activities are determined by adding the reaction mixture containing [^{3}H]SAM plus the carrier DNA directly to a glass fiber filter. For [^{3}H]SAM with a specific activity listed as 2300 Ci/mole, our observed specific activity is 1.7×10^6 cpm/nmole (34% efficiency). Eleven micrograms of completely methylated λ DNA contained about 8000 cpm of tritium or 10 moles of methyl groups incorporated per mole of λ DNA. Reactions used to calculate methylase activities are adjusted so that less than 50% methylation of the substrate takes place in 10 min. One unit is the amount of methylase required to incorporate 1 pmole of CH_3 groups per minute. At the concentration of SAM used in these reactions (1 μM), the reaction rate is at one-half maximum velocity.

A simple and quantitative assay for the *Eco*RI endonuclease employs a circular DNA molecule with one substrate site for the endonuclease. SV40 DNA [6,7], the replicative form of phage G4 (related to ØX174), and the Tc6 [5] *E. coli* plasmid DNA [27,29] are circular DNA molecules known to have one *Eco*RI substrate site. The assay takes advantage of the different mobilities of supercoiled, open circular, and linear forms of DNA during electrophoresis in agarose gels under certain conditions. The gel system is 1.2% agarose in a Tis-borate buffer (10.8 g of Tris base, 0.93 g of Na_2EDTA, and 5.5 g of boric acid per liter). The agarose is melted in the buffer by autoclaving. A vertical slab gel apparatus (a modification of the Reid & Bieleski apparatus [30] that accommodates 14 samples is used. The dimensions of the gel are 3.5 in. x 6 in. x 0.25 in. and the sample wells are made with a lucite comb with 14 teeth, each 0.2 in. wide, and separated by 0.187 in. spaces. A small amount of the molten agarose is applied along the length of the Lucite spacers and allowed to harden before pouring the rest of the molten agarose. After pouring, the sample comb is put in place and the gel allowed to harden. The bottom spacer is removed and replaced with a wetted sponge of the appropriate dimensions.

EcoRI restriction endonuclease reactions are carried out at 37°C for 5 min in 20 μl, containing 100 mM Tris-HCl, pH 7.5, 50 mM NaCl, 5 mM $MgCl_2$, and 0.5 μg SV40 DNA with 4000 to 8000 cpm of ^{32}P/ μg of DNA. Reactions are halted by adding 5 μl of 5% sodium dodecyl sulfate, 25% glycerol plus 0.025% bromophenol blue to the sample well. The gels are run at 175 V for about 1 hr at room temperature or until the dye marker is near the end of the gel. The gel is removed from the mold and soaked in the electrophoresis buffer containing 0.4 μg/ml of ethidium bromide for about 10 to 15 min. When the gel is placed on a long-wave UV lamp (blacklight), the DNA bands can be visualized because of the fluorescence of the intercalated ethidium bromide (see Fig. 1). The bands corresponding to the linear, supercoil and circular forms of SV40 DNA are excised, dehydrated on Whatman #1 filter paper under a heat lamp, and the ^{32}P in each of the forms determined by liquid scintillation counting in Omnitol. The percentage of linear SV40 DNA is used to measure the activity of the EcoRI restriction endonuclease. One unit of activity is defined as the amount of enzyme required to cleave 1 pmole of phospodiester bonds in 1 min.

In Figure 1, a photograph of this gel assay is presented. In this photograph it should be noted that there are two minor bands migrating between the linear monomers of SV40 DNA and the supercoiled forms of the molecule. These represent linear dimer and trimer molecules that have been cleaved once by the EcoRI endonuclease. At excess concentrations of the endonuclease, all of the staining material and ^{32}P are found in the position of the gel corresponding to the linear form of SV40 DNA. For more accurate determinations of the endonuclease activity, the linear dimer and trimer forms are cut out of the gel and counted with the linear monomers. It should also be noted that in these gels the nicked circular forms of the dimers and trimers are present, and they migrate more slowly than the nicked circular form of the monomer.

IV. CULTURE CONDITIONS FOR GROWTH OF BACTERIA

SLBH medium contains, per liter: 11 g of Bactotryptone; 22.5 g of Bacto-yeast extract; 4 ml of glycerol. The pH is adjusted to 7.3 and the medium is autoclaved for 90 min (2-liter aliquots in 6-liter flasks). Phosphate buffer (382.5 ml of 1.0 M K_2HPO_4 plus 117.5 ml of 1.0 M KH_2PO_4 is autoclaved separately for 90 min.

Two Fernbach flasks, each containing 1 liter of SLBH medium plus 0.02% sulfathiazole, are inoculated with a fresh single colony isolate of the RY13 strain and are incubated at 37°C on a rotating platform for 16 hr. The pH of the grown cultures is adjusted to pH 7.3 by the addition of 2.5 N NaOH. These are then combined with 4 liters of fresh SLBH medium and 400 ml of phosphate buffer with or without 60 ml of 2% sulfathiazole in a sterile 18-liter polyethylene carboy of a high-density fermentation unit (Lab-Line/S.M.S. Hi-Density Fermentor, Cat. No. 29500,

Lab-Line Biomedical). When the fermentor is assembled, the carboy is rotated at 375 rpm on a longitudinal axis, partially immersed in a water bath at 32°C. The carboy is flushed with pure O_2 at a flow rate of 8 liters/min and gradually increased to 13 liters/min over a 4-hr period and maintained thereafter at the final rate. Increase in cell mass is monitored on a Klett-Summerson photometer (blue filter). The initial readings are around 185 Klett units. Exponential growth proceeds without a measurable lag with a doubling time of about 90 min if sulfathiazole is present or about 55 min without sulfathiazole. After about 7 hr of incubation, the cell density increases from about 5×10^{10} to 1×10^{11} cells/ml (4200 to 4300 Klett units). During the incubation of the culture, the pH is monitored and kept at pH 7.2 to 7.3. Sixty milliliters of 50% glycerol are added every 2 hr during incubation. The culture abruptly stops increasing in mass around 4200 to 4300 Klett units. Harvesting the 6400 ml of culture at this density yields about 350 g of wet-packed cells. The cells are harvested by low-speed centrifugation in a refrigerated Sorvall RC2-B centrifuge at 4°C (GSA rotor; 10,000 *g*, 15 min). The cell pellets can be kept at 4°C overnight or stored at -20°C for several weeks. Cells can be grown in other fermentation units as well but the yield of cell mass per unit volume is considerably less.

V. SONICATION OF CELLS AND INITIAL FRACTIONATIONS

The cell pellets (350 g) are resuspended in a final volume of 800 ml of extract buffer (EB). EB contains 10 mM KH_2PO_4-K_2HPO_4, pH 7.0, 7 mM 2-mercaptoethanol, and 1 mM EDTA. One hundred milliliter aliquots of the cell suspension are sonicated with a Branson Sonifier (Model LS75) with a 0.5-in. horn at 6 mA for a total of 4 min. The temperature is monitored and not allowed to exceed 12°C. The sonicated cell suspension is centrifuged at 100,000 g (rotor #35 Beckman) for 1 hr in a Beckman preparative ultracentrifuge and the pellet is discarded. The supernatant is adjusted to 1 liter and 350 ml of a freshly prepared 5% (wt/vol) aqueous solution of streptomycin sulfate is added dropwise over a 30-min period and stirred for an additional 30 min. The milky-white solution is centrifuged (GSA Sorvall rotor) for 30 min at 10,000 g and the pellet is discarded. An equal volume (1260 ml) of saturated $(NH_4)_2SO_4$ at 4°C is added to the supernatant over a 30-min period and stirred for an additional ½ hr. The precipitate is centrifuged for 30 min at 10,000 g, the supernatant discarded, and the pellets are dissolved by soaking overnight in a total volume of 300 ml of EB + 0.2 M NaCl and adjusted to final volume of 500 ml EB + 0.2 M NaCl. The EcoRI endonuclease and methylase activities are stable at 4°C under these conditions for at least 2 weeks.

VI. PHOSPHOCELLULOSE CHROMATOGRAPHY

Whatman P11 phosphocellulose is prepared as follows: 125 g of phosphocellulose are stirred into 4 liters of 0.1 N HCl-48% ethanol, and stirred for 30 min at room temperature. The slurry is collected by vacuum filtration on a Buchner funnel and resuspended and stirred into 4 liters of distilled water. This step is repeated until the pH of the slurry is near neutral. The phosphocellulose is collected by filtration and resuspended in 4 liters of 0.1 N NaOH and stirred at room temperature for 30 min. The phosphocellulose is collected again and resuspended in 4 liters of 1 mM EDTA and stirred for 30 min, refiltered, and washed with water until the pH of the slurry is again near neutral. Defining the slurry can be done during the latter steps. The phosphocellulose pH is adjusted to pH 7.0 with 1 N HCl and equilibrated by stirring in EB plus 0.2 M NaCl, pH 7.0. A 4 x 40 cm column of phosphocellulose is constructed at room temperature, transferred to a cold room (4°C), and washed with 2 liter of EB plus 0.2 M NaCl, pH 7.0. It is important to adjust the pH of buffers back to pH 7.0 after the addition of NaCl.

One hundred-milliliter portions of the 50% ammonium sulfate fraction are dialyzed in a BioRad hollow-fiber filter minibeaker (100 ml capacity) by exchanging 4 vol of EB plus 0.2 M NaCl at a flow rate of about 15 to 20 ml/min. Standard dialysis procedures can be used; but the hollow fiber method is faster. The dialyzed fraction is applied to the phosphocellulose column and after adsorption washed with 2 liters of EB plus 0.2 M NaCl. Approximately 10% of the total protein binds to the phosphocellulose. The column is developed with a 4-liter linear NaCl gradient (from 0.2 to 0.8 M) in EB. Fractions of about 27 ml are collected. The OD_{280} profile, NaCl gradient, and EcoRI methylase and endonuclease activities are presented in Fig. 2.

The EcoRI methylase elutes from the phosphocellulose column about 0.4 M NaCl, and the EcoRI endonuclease elutes about 0.6 M NaCl (see Fig. 2). The activity profile of the endonuclease is distributed over two peaks with some trailing of the second peak. The same is found for the methylase activity. This spreading of enzyme activity is reproducible, although the relative amount of activity in the peaks is variable. Our tentative explanation for this spreading is the propensity of these enzymes to aggregate at lower concentrations of NaCl (from 0.1 to 0.2 M) during loading of the column. It is clear, however, that there is separation of the endonuclease and methylase activities at this stage. Fractions 75 to 98 and 116 to 145 were pooled and constitute the phosphocellulose methylase and endonuclease fractions. The phosphocellulose methylase pool has more methylase activity than the $(NH_4)_2SO_4$ fraction. The increase of the methylase activity in the phosphocellulose pool is attributed to the purification of the methylase from a general methylase inhibitor, SAMase, or other enzymes that interact with DNA. The recovery of endonuclease activity in the phosphocellulose endonuclease, on the other hand, is about 30%.

At this and subsequent stages, the methylase and endonuclease fractions will aggregate if the NaCl concentration is reduced below 0.2 M. If precipitation occurs, the precipitated protein can be collected by low-speed centrifugation and resuspended in 0.5 M NaCl + EB, and the activity is partially recovered. The presence of NP40 or Triton X at concentrations of about 0.2% help prevent these proteins from aggregating at lower salt concentrations. If the protein is allowed to precipitate in the presence of NP40, solubilization in EB plus NaCl occurs readily, and the endonuclease has been found to return to its native molecular weight.

VII. HYDROXYAPATITE CONCENTRATION OF THE *Eco*RI ENDONUCLEASE

Hydroxyapatite (obtained from Clarkson Chemical Co.) is equilibrated in EB, 0.2 M NaCl, 0.2% NP40, and columns with 10-ml bed volumes are made and washed with 100 ml of the same buffer. The pooled endonuclease and methylase fractions are applied directly to a hydroxyapatite column and washed with 5 to 10 ml of EB, 0.2 M NaCl, 0.2% NP40; the activities are eluted with 500 mM KH_2PO_4-K_2HPO_4, pH 7.0, 0.2 M NaCl, 0.2% NP 40, 7 mM 2-mercaptoethanol, 1 mM EDTA. The endonuclease and methylase can be stored in this buffer after elution.

VIII. DEAE-CELLULOSE CHROMATOGRAPHY

Whatman DE-52 DEAE-cellulose is prepared as described in the Whatman information leaflet, I-L2. The DEAE-cellulose slurry is equilibrated at room temperature in TB buffer (20 mM Tris-HCl, pH 7.5, 0.2% NP40, 7 mM 2-mercaptoethanol, 1 mM EDTA, 0.05 M NaCl), and a 2 x 20 cm column is constructed. The hydroxyapatite-concentrated endonuclease is dialyzed against TB buffer that has a NaCl concentration of 0.2 M. The dialyzed fraction is diluted fourfold with TB buffer and immediately applied to the DEAE-cellulose column. About 5-ml volumes of the fraction are applied at a time in order to minimize the length of time the endonuclease is in a low salt buffer. After adsorption of the enzyme, the column is washed with 100 ml of TB buffer and then a linear gradient of NaCl (0 to 0.5 M) in TB buffer is applied to the column and 7.7-ml fractions are collected. The endonuclease activity elutes about 0.1 M NaCl. As can be seen in Fig. 3, the major protein peak does not correspond to the peak of *EcoRI* endonuclease activity. A shallower NaCl gradient should resolve the major contaminating protein peak from the endonucleolytic activity. At this stage in the purification, the low protein concentration and possibly the low NaCl concentration become important factors in the stability of the endonuclease. Therefore, the main DEAE-cellulose fractions (25 to 31 in this case) and side fractions are pooled as soon as possible and concentrated as described in Sect. VII. The pooled and concentrated DEAE-cellulose fraction of endonuclease activity is purified about fourfold over the phosphocellulose pool, and purified 180-fold compared with the $(NH_4)_2SO_4$ fraction (Table 2). Comparison of the various fractions by SDS-acrylamide gel electrophoresis reveals that the main DEAE-cellulose fraction has two detectable proteins (Fig. 4). The major staining protein has a molecular weight probably greater than 100,000 when compared with the standard proteins. The second

TABLE 2

Purification of *Eco*RI Endonuclease

Step	Total protein[a]	Endonuclease total units	Specific activity	Purification	Recovery, %
A. 0-50% $(NH4)_2SO_4$	8550 mg	132,000	15	--	100
B. Phosphocellulose pool	56.4	37,700	668	45	28
C. Hydroxyapatite concentrate	26.2	43,500	1660	110	33
D. Dialyzed hydroxy-apatite concentrate	23.8	30,400	1280	85	23
E. DEAE cellulose pool	6.5	11,100	1710	114	8
F. Hydroxyapatite concentrate	4 mg	11,100	2780	185	8
G. Sephadex pool[b]	< 200 µg	4,600	$> 23,000$	> 1500	3.5

[a]Protein concentrations were measured by a modified procedure of Lowry [31]. The presence of NP40 detergent in the Lowry reaction results in a massive precipitate that is removed by low-speed centrifugation.

[b]Calculated from the recoveries of protein and endonuclease units from 1/6 of the hydroxy-apatite-concentrate sample applied to the Sephadex column.

protein found in this fraction is much smaller, having a molecular weight around 30,000, suggesting that this is the subunit of the EcoRI restriction endonuclease, which has a native molecular weight of about 60,000.

At this stage of purification the endonuclease should be diluted for assaying activity in EB, 0.2 M NaCl, 0.2% NP40, and gelatin at a concentration of 100 μg/ml. Gelatin should also be included in the assay reaction at this and subsequent stages to minimize inactivation of the endonuclease because of the dilute protein concentrations.

IX. SEPHADEX GEL FILTRATION

The concentrated DEAE-cellulose fraction can be further purified by gel filtration on Sephadex G-100. A 2 x 45 cm G-100 Sephadex column equilibrated in NB buffer (10 mM Na_2HPO_4-NaH_2PO_4, pH 7.0, 7 mM 2-mercaptoethanol, 1 mM EDTA, 0.2% NP40, 0.2 M NaCl) is constructed, and a 1.0-ml sample of the concentrated DEAE-cellulose fraction is applied to the column. About 90% of the total protein applied is found in the void volume, while the EcoRI endonuclease is retarded (Fig. 5). The protein in the void volume is the high-molecular-weight protein component of the DEAE-cellulose fraction. The pooled Sephadex fractions with the EcoRI endonuclease activity has the lower-molecular-weight component of the DEAE-cellulose fraction (Fig. 4).

X. DISCUSSION

The EcoRI endonuclease can be purified to near homogeneity (as assessed by SDS-polyacrylamide gel electrophoresis) by the procedures described here. The concentrated phosphocellulose fraction (step C) is free of significant nonspecific double-strand endonucleases and can be used for the purpose of preparing defined DNA fragments from susceptible molecules. This and other fractions are usually dialyzed against phosphate (10 mM) and stored at 4°C. The presence of 0.2 M NaCl and 0.2% NP40 in the storage buffer keeps the EcoRI endonuclease from aggregating.

Electron microscopic observation shows that linear SV40 DNA molecules made with step C enzyme are capable of forming circular and linear concatemers at 4°C, suggesting that the cohesive termini are intact and that this fraction is free of any significant exonuclease that would destroy these termini. However, subsequent fractions of the endonuclease generate linear SV40 DNA molecules that appear to reassociate their cohesive termini more readily.

Some of the more pertinent parameters of the EcoRI endonuclease reaction are: it has a broad Mg^{2+} optimum (from 1 to 15 mM); it is stimulated two- to threefold by NaCl concentrations from 50 to 100 mM and inhibited by NaCl above concentrations of 100 mM; gelatin is present in the diluent; and the reaction protects the purified endonuclease.

The reaction appears to be first order, and the rate is therefore dependent on the amount of substrate. At SV40 DNA concentrations of 0.05 to 0.1 μg/μl (from 5 to 10 μg total) with 0.3 units of EcoRI endonuclease, the reaction goes to completion (95% linear DNA products) in 10 min. The EcoRI endonuclease can be inactivated by incubation at 62°C for 5 min or by the addition of SDS or EDTA.

The EcoRI methylase can be used to calculate the number of EcoRI endonuclease substrate sites on any given molecule by measuring the incorporation of CH_3 groups into the DNA of interest and dividing by two. The methylase can also be used to determine how many EcoRI sites have been cleaved since the EcoRI methylase cannot methylate the cleaved substrate.

ACKNOWLEDGMENTS

This investigation was supported by Public Health Service Grants GM 14378, AI 00299, and CA 14026; Cancer Research Funds of the University of California; University of California MSC #30 Hampton Fund; and American Cancer Society Grant #NP-112A. We are grateful to Dr. Louise Chow for the electron microscopic analysis of the SV40 DNA.

REFERENCES

1. K. J. Danna and D. Nathans, Proc. Natl. Acad. Sci., U.S., 68, 2913 (1971).

2. D. Nathans and K. J. Danna, Nature New Biol., 236, 200 (1972).

3. D. Nathans and K. J. Danna, J. Mol. Biol., 64, 515 (1972).

4. K. J. Danna and K. Nathans, Proc. Natl. Acad. Sci., U.S., 69, 3097 (1972).

5. K. J. Danna, G. H. Sack, and D. Nathans, J. Mol. Biol. (1973), in press.

6. J. Morrow and P. Berg, Proc. Natl. Acad. Sci., U.S., 69, 3365 (1972).

7. H. Delius and C. Mulder, Proc. Natl. Acad. Sci., U.S., 69, 3215 (1972).

8. C. C. Fareed, G. F. Garon, and N. P. Salzman, J. Virol., 10, 484 (1972).

9. J. Morrow, P. Berg, T. J. Kelly, and A. M. Lewis, J. Virol., (1973), in press.

10. C. A. Hutchison and M. H. Edgell, J. Virol., 8, 181 (1971).

11. H. W. Boyer, Ann. Rev. Microbiol., 25, 153 (1971).

12. J. A. Lautenberger and S. Linn, J. Biol. Chem., 247, 6176 (1972).

13. B. Eskin and S. Linn, J. Biol. Chem., 247, 6183 (1972).

14. M. Meselson, R. Yuan, and J. Heywood, Ann. Rev. Biochem., 447, 41 (1972).

15. H. O. Smith and K. Wilcox, J. Mol. Biol., 51, 379 (1970).

16. R. N. Yoshimori, Ph.D. Thesis, University of California, San Francisco, 1971.

17. R. Gromokava and S. H. Goodgal, J. Bacteriol., 109, 987 (1972).

18. T. J. Kelly and H. O. Smith, J. Mol. Biol., 51, 393 (1970).

19. J. Hedgpeth, H. M. Goodman, and H. W. Boyer, Proc. Natl. Acad. Sci. U.S., 69, 3448 (1972).

20. J. E. Mertz and R. W. Davis, Proc. Natl. Acad. Sci. U.S., 69, 3370 (1972).

21. H. W. Boyer, L. T. Chow, A. Dugaiczyk, J. Hedgpeth, and H. M. Goodman, Nature New Biol. (London), 244, 40 (1973).

22. A. Dugaiczyk, J. Hedgpeth, H. W. Boyer and H. M. Goodman, Biochemistry, 13, 503 (1974).

23. S. P. Adler and D. Nathans, Federation Proc. (Abstr.), 29, 725 (1970).

24. K. Horiuchi and N. D. Zinder, Proc. Natl. Acad. Sci. U.S., 69, 3220 (1972).

25. R. Davis, personal communication, 1973.

26. R. B. Helling, H. M. Goodman, and H. W. Boyer, J. Virol. (1974), in press.

27. N. Godson and H. W. Boyer, J. Virol. (1974), in press.

28. R. N. Yoshimori, D. Roulland-Dussoix, and H. W. Boyer, J. Bacteriol., 112, 1275 (1972).

29. S. N. Cohen, A. C. Y. Chang., Proc. Natl. Acad. Sci. U.S., 70, 1295 (1973).

30. J. V. Maizel, in *Methods in Virology* (K. Maramorosch and H. Koproski, eds.), Vol. 5, Academic Press, New York, 1971, p. 180.

31. J. Leggett-Bailey, in *Techniques in Protein Chemistry*, 2nd ed., Elsevier, Amsterdam and New York, 1967, p. 340.

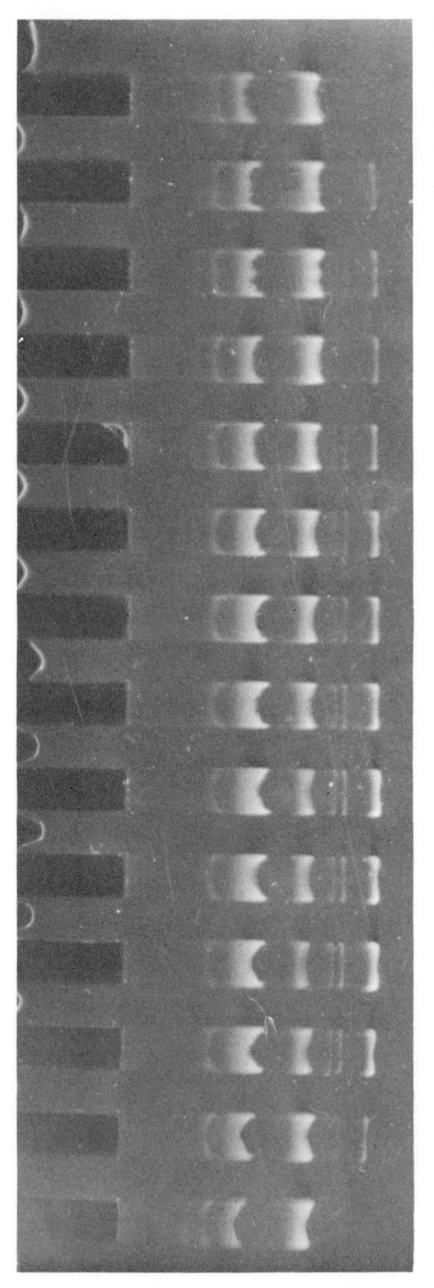

FIG. 1. Agarose gel assay for EcoRI endonuclease. The gel and assays are described in detail in Section V. The gel was placed on a blacklight and photographed with a Polaroid Camera with a #9 Wratten filter covering the lens (Type 55 film). The linear monomeric form of SV40 DNA is the fastest migrating component (migration is downward). The two major, rather broad bands are super-coiled and nicked circular DNA, with the latter migrating more slowly. The two minor bands migrating behind the linear SV40 DNA are linear dimer and trimer molecules of SV40 DNA. The sample well on the left of the gel contained SV40 DNA without enzyme. Note the two slowest migrating components that represent nicked circular dimer and trimer molecules. This gel represents the assays of the sephadex column of Section IX. The percentage of ^{32}P in the linear forms for the various Sephadex fractions are from left to right: control, 4.6%; 29, 6.9%; 30, 21.1%; 31, 34.1%; 32, 25.1%; 33, 25.1%; 34, 17.1%; 35, 13.3%; 36, 12.4%; 37, 8.2%; 38, 7.0%; 39, 6.8%; 40, 6.1%; control, 3.5%.

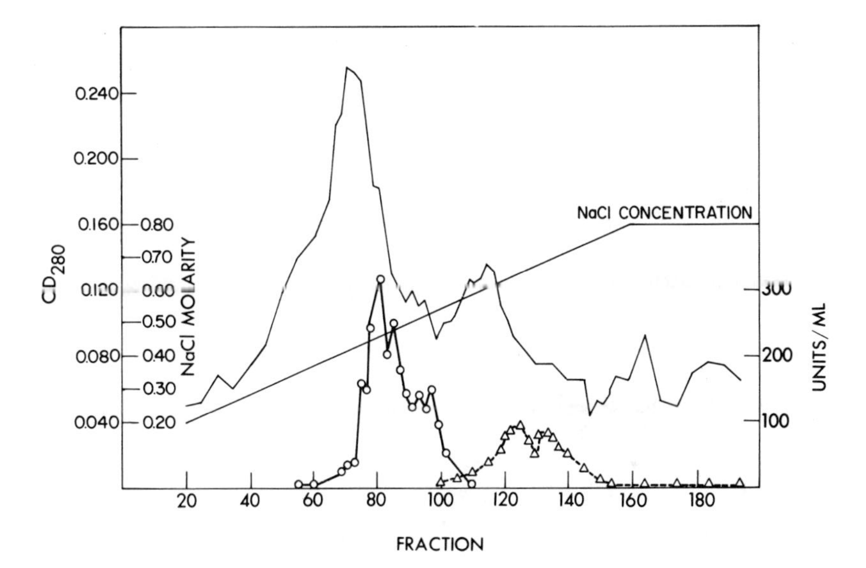

FIG. 2. Phosphocellulose chromatography of the EcoRI methylase and endonuclease. The details for loading of the sample of the elution conditions are described in Section VI. The solid line represents OD_{280}, the open circle represents the EcoRI methylase activity, and the open triangles represent the EcoRI endonuclease activity.

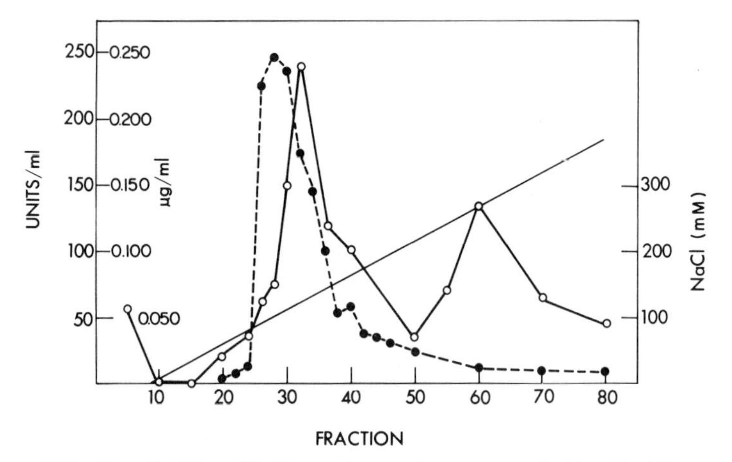

FIG. 3. DEAE-cellulose chromatography of the EcoRI endonuclease. Details for loading of the sample and elution condition are described in Section VII. Since NP40 absorbs strongly at 280 nm, OD profiles cannot be used. Protein concentrations were determined as described in Table 2. The closed circles represent EcoRI endonuclease activity and the open circles represent micrograms per milliliter of protein.

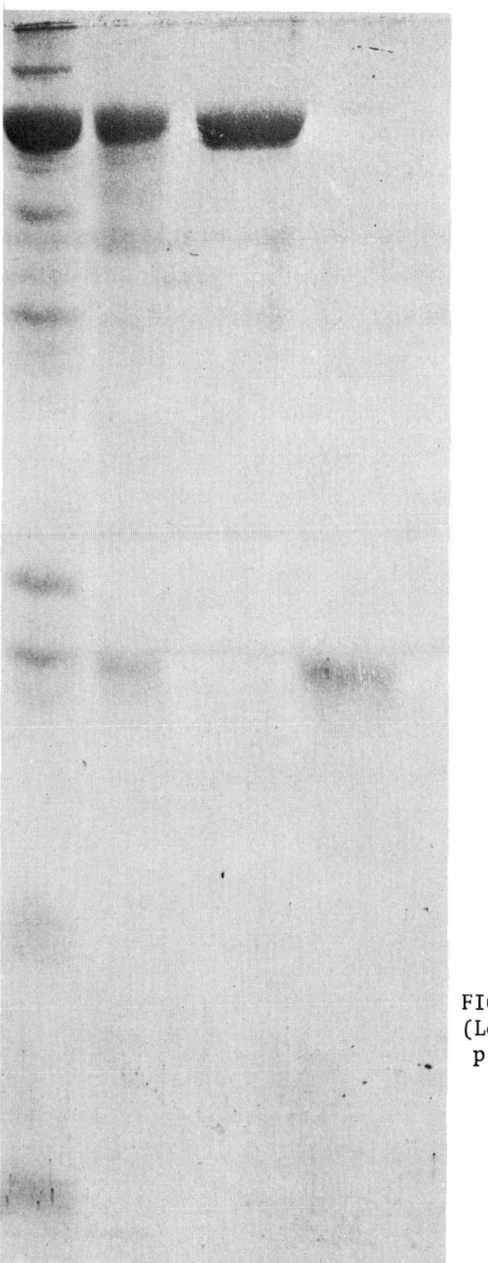

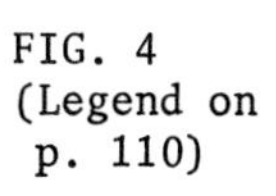

FIG. 4
(Legend on p. 110)

FIG. 4. SDS-polyacrylamide gel electrophoresis. The samples are heated at 100°C in 1% SDS and 1% 2-mercaptoethanol prior to application. The discontinuous buffer system is that developed by Laemmli and Maizel [30] with a separating gel of 10% acrylamide, 0.4% bisacrylamide. The samples are electrophoresed at 50 V for 30 min, then 100 V for 3.5 hr at room temperature. The samples shown from left to right are (a) endonuclease phosphocellulose pool (fraction D, Table 2); (b) DEAE fractions 25-31; (c) Sephadex G-100 fraction 21 (excluded volume); and (d) Sephadex G-100 fractions 30-34.

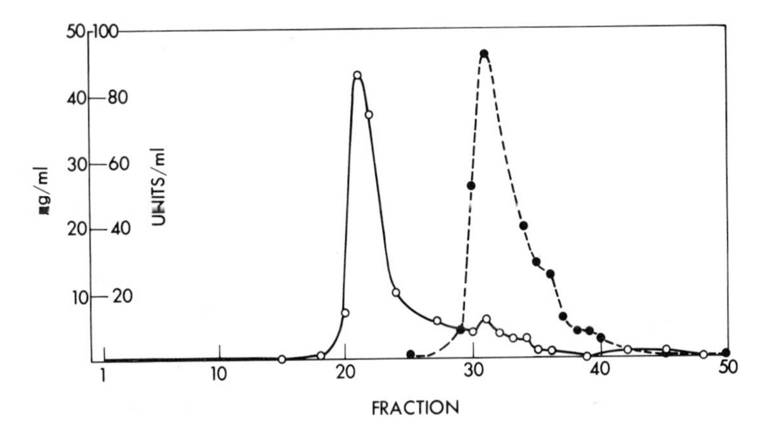

FIG. 5. Sephadex G-100 gel filtration. A 1.0 ml sample of the DEAE hydroxyapatite concentrate was applied to the Sephadex G-100 column described in detail in Section IX. Fractions of 2.4 ml were collected. The open circles represent microgram per milliliters of protein and the closed circles represent EcoRI endonuclease activity.

Chapter 5

RESTRICTION ENDONUCLEASES AP, GA, AND H-I FROM THREE HAEMOPHILUS STRAINS

M. Takanami

Institute for Chemical Research
Kyoto University
Uji, Kyoto, Japan

I. INTRODUCTION

Restriction endonucleases, which specifically degrade foreign DNA at a limited number of sites, have been isolated from several bacterial strains [1-7] and used to obtain unique fragments from viral DNA [8-11]. This type of enzyme appears to produce duplex cleavages on double-stranded DNA by recognizing specific nucleotide sequences [6,12]. Three species of enzymes with different cleavage site specificities were recently isolated from three *Haemophilus* strains; *H. aphirophilus*, *H. gallinarum*, and *H. influenzae* H-I (abbreviated as Endo AP, GA, and H-I) [13,14]. The respective enzymes produce different sizes of fragments from phage DNA. From the analysis of terminal nucleotide sequences, Endo AP was shown to cleave DNA at a sequence of four nucleotide pairs with a twofold rotational symmetry [15]. The great importance of these enzymes for elucidating the structure and biologic function of DNA has been emphasized [8].

II. PREPARATION OF ENZYMES

A. Strains

Strains, their sources, and abbreviations used for enzymes are given in Table 1. These strains form translucent colonies on brain-heart-infusion agar (Difco) supplemented with NAD (2 μg/ml) and hemin (10 μg/ml). Strains are maintained either by transfer to fresh plates every two weeks or by freezing (below -60°C) in the presence of 30% glycerol.

TABLE 1

Strains and Abbreviations Used for Enzymes

Enzyme	Strains
Endo AP	*Haemophilus aphirophilus* (ATCC 19415)
Endo GA	*Haemophilus gallinarum* (NCTC 3438)
Endo H-I	*Haemophilus influenzae* H-I[a]

[a]A strain isolated at Research Institute for Microbial Diseases, Osaka University.

B. Growth of Cells

All strains grow well in a medium, which contains 35 g brain-heart infusion (Difco), 2 mg NAD, and 10 mg hemin in 1 liter. For preparation of the medium, brain-heart infusion is separately autoclaved; after cooling, stock solutions of NAD and hemin are added to the indicated amounts. The NAD solution (4 mg/ml) is sterilized by passage through a millipore filter. Hemin is first dissolved in 0.5% NH_4OH to a concentration of 10 mg/ml, followed by the addition of ethanol to 45%. These solutions may be stored for at least one month at 4°C. Cells are grown with vigorous shaking or with aeration by a bubbler. Approximate division time is 40 min. When the cell density reaches about OD_{600} = 1, cells are harvested by centrifugation, washed once with saline, and stored in a freezer (below -20°C). The yield from a 10-liter culture is about 20 g (wet weight).

C. Preparative Procedure

Frozen cells (about 20 g) are suspended in 50 ml of 0.05 M Tris(pH 7.6)-1 mM mercaptoethanol, disrupted by a sonicator (e.g., Raytheon Sonic Oscillator) for 5 min at 6 A output, and centrifuged for 30 min at 40,000 rpm in the Spinco 40 rotor. Ammonium sulfate is added to the supernatant to obtain a fraction from 35 to 65% saturation. The precipitate is dissolved in 50 ml of 10 mM potassium phosphate buffer (pH 7.6)-1 mM mercaptoethanol (buffer A), and dialyzed against 2 liters of the same buffer for 3 to 4 hr.

The dialyzed fraction is applied to a phosphocellulose (Whatman P-11) column (1.8 cm x 12 cm) and equilibrated with buffer A. After washing the column with 100 ml of buffer A, adsorbed proteins are eluted with a linear gradient of KCl (buffer A to buffer A + 1 M KCl, 150 ml each). Eight milliliter fractions are collected, and aliquots (5 to 30 μl) of fractions are added to assay mixtures containing phage fd replicative form I DNA(fd RF I DNA). After incubation for 30 min at 37°C, the fd RF I DNA remaining is assayed by transfection (Section III,A). The activity peak regions are collected and dialyzed against 2 liters of buffer A for 3 to 4 hr. For routine purposes, the transfection assay can be omitted, and the approximate fractions containing the enzyme are collected without assay (see Table 2).

The dialyzed fraction is charged onto a DEAE-cellulose (Whatman DE-52) column (1 cm x 12 cm) equilibrated with buffer A. After washing the column with 30 ml of buffer A, adsorbed proteins are eluted with a linear gradient of KCl (buffer A to buffer A + 0.4 M KCl, 100 ml each). From these, 5-ml fractions are collected and their activity is assayed by transfection (Section III,A). The specificity of active fractions is then assayed by the gel electrophoresis method 1 (Section III,B,1).

For routine purposes, the specificity of the approximate fractions containing the enzyme can be determined without the transfection assay.

Typical elution patterns of Endo GA from P-11 and DE-52 columns are shown in Fig. 1. The chromatographic profiles are different with varied strains. Approximate KCl concentrations eluting enzyme from P-11 and DE-52 columns are given in Table 2.

TABLE 2

Approximate KCl Concentrations Eluting Enzymes from Whatman P-11 and DE-52 Columns

Enzyme	P-11 column	DE-52 column
Endo AP	0.24-0.28 M	0.07-0.09 M
Endo GA	0.34-0.38 M	0.05-0.07 M
Endo H-I	0.28-0.32 M	0.11-0.13 M

The fractions that yield discrete bands of DNA fragments are collected and charged onto a P-11 column (0.8 cm x 10 cm) and equilibrated with buffer A. Adsorbed proteins are eluted with a linear gradient from 0.1 to 0.6 M KCl in buffer A (50 ml each). Three milliliter fractions are collected and the regions containing enzyme (see Table 2) are assayed by gel electrophoresis method 2 (Section III, B,2). Fractions that yield sharp bands against a clear background are collected, concentrated by dialysis against buffer A containing 50% glycerol, and stored at -20°C.

Crude cell extracts contain some nonspecific nucleases, the bulk of which are removed by the initial P-11 column chromatography. On assay of the enzyme specificity by gel electrophoresis method 2 (Section III,B,2), contamination by such nonspecific nucleases results in radioautographs with a smeared background, instead of sharp bands against a clear background. The contaminating nucleases are removed by the second P-11 column. A sephadex G-200 column chromatography is also applicable for this purpose. In the preparation of Endo AP, fractions from the DE-52 column can usually be used without further purification because of the low content of such nonspecific nucleases in the original extract.

Table 3 presents the purification data of Endo GA. The yield of Endo H-I is about equal to that of Endo GA. In contrast, about three times as much Endo AP can be prepared starting from about an equal weight of cells.

TABLE 3

Purification of Endo GA

Step	Total protein, mg	Specific activity, units/mg
Crude extract	2654	--
Ammonium sulfate fraction (35%-65%)	1652	1.3
Phosphocellulose chromatography	15.2	76
DEAE-cellulose chromatography	0.51	2060
Re-chromatography on phosphocellulose	0.26	2480

III. ASSAY OF ENZYME ACTIVITY

A. Assay by Transfection

1. Preparation of Lysozyme Spheroplasts

E. coli K38 is grown in 300 ml of peptone broth to a cell density of 10^9/ml and harvested by centrifugation. Cells are suspended in 15 ml of 0.6 M sucrose-0.04 M Tris (pH 8.5). After 15 min at 37°C, 0.25 ml of 0.2 M EDTA, 1.5 ml of 10% serum albumin, and 0.1 ml of 2 mg/ml lysozyme are added. The mixture is held for 1 min at 20°C and immediately chilled; 0.2 ml of 1 M $MgSO_4$ is added to inhibit the lysozyme action. The spheroplasts may be stored in a refrigerator for at least one week without loss of activity.

2. Assay Procedure

Digestion is carried out in a reaction mixture containing 10 mM Tris (pH 7.6), 7 mM $MgCl_2$, 7 mM mercaptoethanol, 0.01 OD_{260} unit of fd RF I DNA, and enzyme, in a final volume of 0.3 ml. After incubation at 37°C, 0.6 ml of dilution buffer (0.1 M NaCl-5 mM $MgSO_4$-1 mM $CaCl_2$-0.1 M sucrose) and 0.1 ml of lysozyme spheroplasts are added. The mixture is held for 5 min at 37°C, then mixed with 0.1 ml of fresh *E. coli* K38 cells ($\sim 10^8$) and 2 ml of soft agar containing 0.4 M sucrose, 5 mM $MgSO_4$, and 1 mM $CaCl_2$, and plated in the usual manner. Under these conditions, about 500 plaques per plate are usually formed without enzymatic digestion. From fd-infected *E. coli* K38 cells, fd RF I DNA is prepared as described previously [16].

This method is used for a primary survey of enzyme activity in chromatographic fractions.

3. Definition of Enzyme Unit

An enzyme activity unit is expressed as that activity that destroys the infectivity of 0.01 OD_{260} unit of fd RF I DNA within 30 min at 37°C.

B. Assay by Polyacrylamide Gel Electrophoresis

1. Method 1

The composition and volume of the reaction mixture are the same as those used for the transfection assay, except that 0.3 OD_{260} unit of fd RF I DNA is added. Since other phage DNA such as T3, Ø80, and λDNA all give characteristic patterns of fragments, fd RF I DNA can be replaced by these DNA species. The reaction is usually carried out for 2 to 3 hr at 37°C, and terminated by adding 20 μl of 0.2 M EDTA, followed by addition of 30 μl of 40% sucrose. The mixture is layered on a 5% or 10% gel column (0.6 cm x 12 cm) formed in 0.036 M Tris-0.032 M KH_2PO_4-1 mM EDTA (pH 7.8) (running buffer). Electrophoresis is usually carried out for 16 hr at 2 mA/tube and at room temperature. Gels are removed, stained for 1 hr with 0.4% acridine orange dissolved in 1 N acetic acid, and then destained with 0.2 N acetic acid. Roughly, 0.002 of an OD_{260} unit of a single DNA fragment species is visible as a stained band.

A routine procedure for preparation of eight 5% gel columns is as follows. Five milliliters of 30% acrylamide (acrylamide : methyl-bis-acrylamide = 30 : 1), 15 ml of double-strength running buffer, 9.45 ml of distilled water, 0.05 ml of tetramethylethylenediamine, and 0.5 ml of freshly prepared 5% ammonium persulfate are mixed in this order. The solution is poured into columns, and running buffer is layered on the top of the columns to obtain a flat surface.

This method is used for the early purification stage to find fractions containing specific enzymes.

2. Method 2

The reaction conditions are similar to those used in other assay methods, except that [^{32}P] DNA with high specific activity ($10^4 \sim 10^5$ cpm/0.01 OD_{260} unit) is used as substrate. Incubation is carried out for more than 6 hr at 37°C in the presence of an excess of enzyme. After electrophoresis, gels are split vertically into halves, covered with thin plastic films, and exposed to X-ray films.

This method is used to obtain complete digestion patterns of a DNA molecule and also to examine the purity of enzyme preparations. The radioautographs of digestion products will give sharp bands against a clear background even after incubation for a long period if the enzyme preparation does not contain any other nonspecific nuclease.

IV. PROPERTIES OF RESTRICTION ENDONUCLEASES

The enzyme activity is strikingly dependent on the presence of Mg ions. The optimal concentration is around 5 mM for all enzymes (Fig. 2). Mg ions can be replaced by Mn ions. The specificity of these enzymes is not altered by this substitution, as judged by gel electrophoresis of digestion products. The maximum activity is obtained at around pH 7 to 8 (Fig. 2). The enzyme activity is not influenced by salts (0 to 0.5 M KCl or NaCl). All enzymes are stable for at least eight months when stored at -20°C in the presence of 50% glycerol.

V. CLEAVAGE SITE SPECIFICITIES OF ENZYMES

A. Specific Cleavage of Phage DNA

Since an incomplete digestion yields many intermediate fragments, the number of DNA fragments produced by specific cleavage should be determined by exhaustive digestion of DNA. Figure 3 demonstrates that an incomplete digestion of fd RF I DNA by Endo H-I yields seven discrete fragments, whereas only three fragments are obtained after the complete digestion. The appearance of seven fragments at an early stage of digestion is a proof that the number of cleavage sites is three, as four partially-digested fragments (1, 2, 2, and 2 cleavages) would be anticipated.

The number of cleavage sites on a DNA molecule can be determined by taking radioautographs of the DNA fragments produced from highly labeled [^{32}P] DNA by a long period of incubation with an excess of enzyme. Figure 4 shows radioautographs of fragments produced from RF I [^{32}P] DNA of fd by cleavage with the three enzymes described here and two other Haemophilus enzymes, Endo R [2] and Z [4]. These electrophoretic patterns are highly reproducible and are not altered by further incubation. The number of fragments produced by these enzymes have been determined from the distribution of radioactivity in the bands [14]. Endo H-I produces only three fragments, as mentioned above, Endo GA produces six fragments, and Endo AP 13 fragments altogether. Endo R cleaves this DNA molecule at one site and Endo Z at 11 sites, respectively. These observations imply that the cleavage site specificities of the enzymes from the five Haemophilus strains are all different.

These enzymes also cleave other phage DNAs, such as T3 and λ DNA, into a large number of fragments. Upon resolution of fragments on gel electrophoresis, the resulting patterns are different depending on the species of DNA and the enzyme (Fig. 5).

As has been shown with Endo R [12] and RI [6], restriction endonucleases appear to induce duplex cleavages on DNA by recognizing specific nucleotide sequences. It is therefore likely that DNA can be cleaved into even smaller fragments by a combination of enzymes with different cleavage site specificities. This was clearly demonstrated by the cleavage of fd RF I DNA [14]. Endo R specifically cleaved one of the fragments produced by other Haemophilus enzymes into two pieces. By other combinations such as indicated in Fig. 4, many smaller fragments were produced that were not seen in the digests of the individual enzymes. Although the exact number of fragments were not counted because of overlapping, these patterns were very reproducible. DNA treated by two different enzymes sequentially gave a pattern identical with that treated by a mixture of these enzymes. After isolating a particular fragment from a digest, therefore, the fragment can be cleaved into smaller pieces by other enzymes.

B. Nucleotide Sequence Restricted by Endonuclease AP

Terminal nucleotide sequences of fragments produced from fd RF I DNA and T3 DNA by cleavage with Endo AP were recently determined [15]. The results of these analyses are briefly summarized below.

DNA fragments produced by Endo AP received ^{32}P from [γ ^{32}P] ATP in the polynucleotide kinase reaction only after treatment with alkaline phosphatase, indicating that the fragments have 5' phosphoryl termini. The 5'(^{32}P) fragments were hydrolyzed to mononucleotides by pancreatic DNase and venom phosphodiesterase. ^{32}P was only recovered at 5'-CMP. Di-, tri-, and tetranucleotides containing ^{32}P at the 5' termini were prepared from the 5'(^{32}P) fragments, and resolved by Dowex I column chromatography. The dinucleotide fraction yielded a ^{32}P peak only at the pCpG region and the trinucleotide fraction at the pCpGpG region. In contrast, the tetranucleotide fraction produced four ^{32}P peaks. It was concluded that both of the 5'-termini fragments had the sequence of pCpGpGpN----(N = A,C,G, or T).

For the determination of the 3' termini sequences, uniformly labeled [^{32}P] DNA was hydrolyzed by Endo AP. Following hydrolysis by micrococcal nuclease, the NpN fraction from the 3' termini was isolated. Digestion of this fraction by venom phosphodiesterase yielded only 5'-CMP and digestion by spleen phosphodiesterase yielded all four 3' mononucleotides, indicating that the 3'-termini of both fragments had the sequence of ---NpC. With these observations, it was concluded that Endo AP induced duplex cleavages on

DNA at the following sequences:

$$5'\text{-------NpC}\overset{\downarrow}{\ }\text{pCpGpGpN-------}3'$$
$$3'\text{-------NpGpGpCp}\underset{\uparrow}{\ }\text{CpN-------}5'$$

The sequence has a twofold rotational symmetry, in accord with the sequences for Endo R and RI. However, this enzyme restricts only a sequence of four nucleotides, whereas Endo R and RI cleave DNA at unique hexa- and octanucleotide sequences, respectively. Consistent with this specificity that Endo AP cleaves λ DNA at a great number of sites (Fig. 5) in contrast to Endo RI, which cleaves this DNA molecule into only six discrete fragments [11]. Nucleotide sequences restricted by Endo GA and H-I have not been determined yet, but these enzymes cleave T3 and λ DNA into 40 to 60 fragments (Fig. 5). It is therefore likely that these enzymes also restrict relatively short nucleotide sequences. As shown with fd RF I DNA, the number of cleavage sites on DNA increase additvely by the combination of enzymes with different cleavage site specificities. The enzymes described here would be particularly useful for obtaining DNA segments short enough for the determination of nucleotide sequences.

Note added in proof

According to the nomenclature proposed by H. O. Smith and D. Nathans (J. Mol. Biol., 81, 419, 1973), restriction endonucleases AP, GA, and H-I should be named as endonucleases R·Hap, R·Hga and R·HinH-I, respectively.

REFERENCES

1. M. Meselson and R. Yuan, Nature, 217, 110 (1968).

2. H. O. Smith and K. W. Wilcox, J. Mol. Biol., 51, 379 (1970).

3. R. Gromkova and S. H. Goodgal, J. Bacteriol., 109, 987 (1972).

4. J. H. Middleton, M. H. Edgell, and C. A. Hutchison, III, J. Virol., 10, 42 (1972).

5. B. Eskin and S. Linn, J. Biol. Chem., 247, 6183 (1972).

6. J. Hedgpeth, H. M. Goodman, and H. W. Boyer, Proc. Natl. Acad. Sci. U.S., 69, 3448 (1972).

7. P. A. Sharp, B. Sugden, and J. Sambrook, Biochemistry, 12, 3055 (1973).

8. K. J. Danna and D. Nathans, Proc. Natl. Acad. Sci. U.S., 68, 2913 (1971).

9. J. F. Morrow and P. Berg, Proc. Natl. Acad. Sci. U.S., 69, 3365 (1972).

10. U. Pettersson, C. Mulder, H. Delius, and P. A. Sharp, Proc. Natl. Acad. Sci. U.S., 70, 200 (1973).

11. A. Allet, P. G. N. Jeppesen, K. J. Katagiri, and H. Delius, Nature, 241, 120 (1973).

12. T. J. Kelly, Jr., and H. O. Smith, J. Mol. Biol., 51, 393 (1970).

13. M. Takanami and H. Kojo, FEBS Letters, 29, 267 (1973).

14. M. Takanami, FEBS Letters, 34, 318 (1973).

15. H. Sugisaki and M. Takanami, Nature (New Biology), 246, 153 (1973).

16. M. Sugiura, T. Okamoto, and M. Takanami, J. Mol. Biol., 43, 299 (1969).

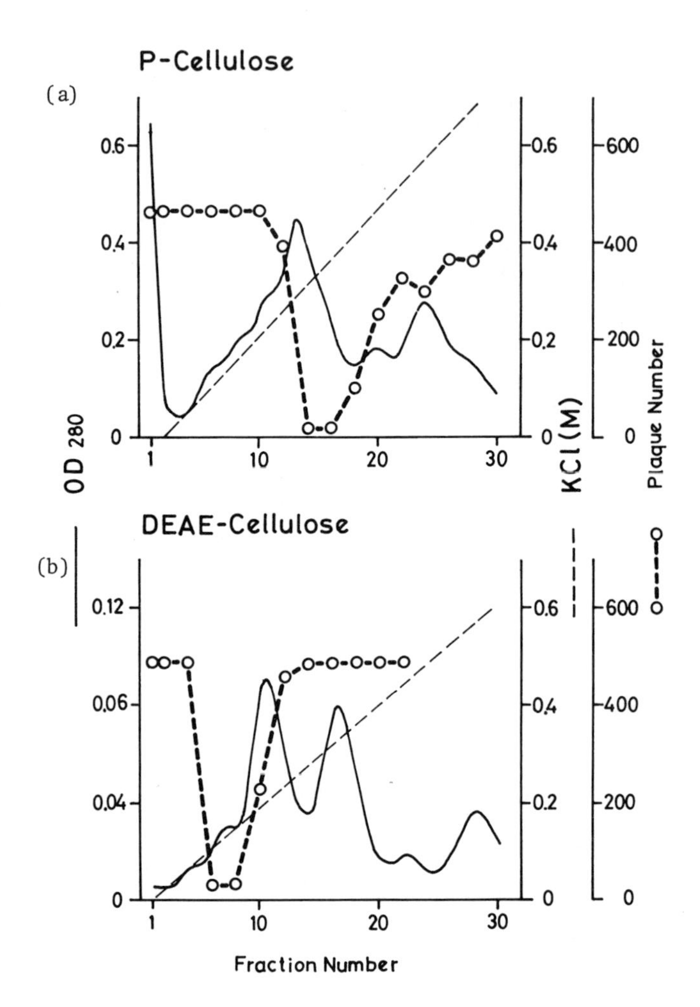

FIG. 1. (a)Purification of Endo GA by chromatography on phospho- and DEAE-cellulose columns. The 35% to 65% ammonium sulfate fraction from the extract of H. gallinarum was loaded on a Whatman P-11 column and eluted with a linear gradient of KCl. (b)Fractions destroying the infectivity of fd RF I DNA (14th to 17th tubes) were then chromatographed on a Whatman DE-52 column.

(c)

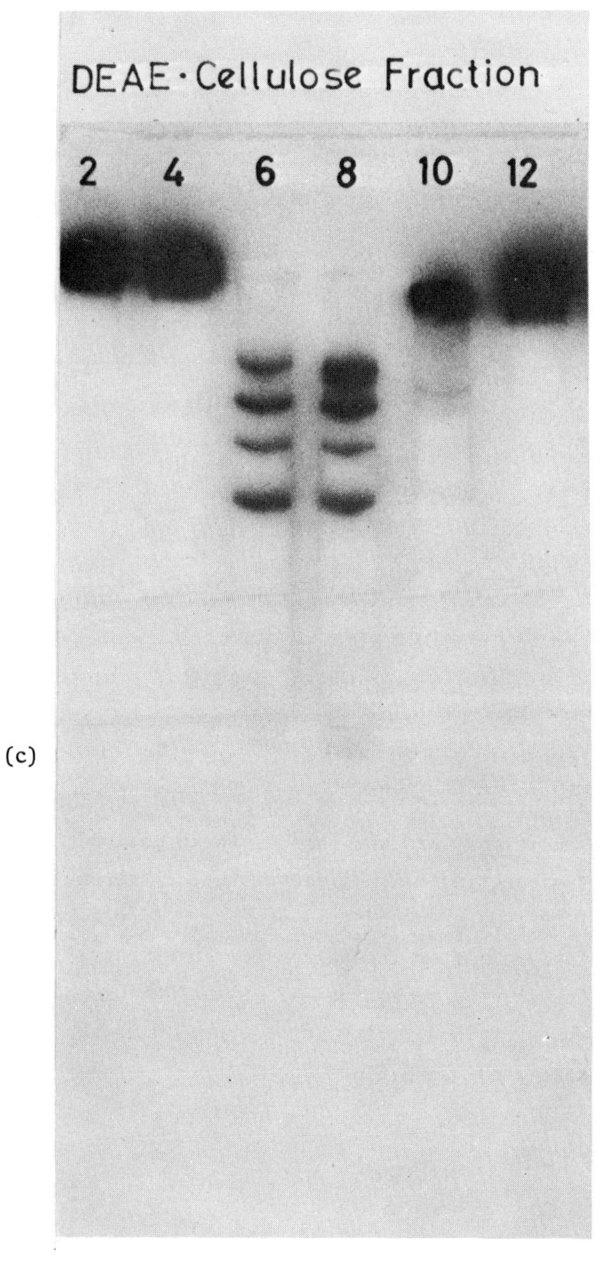

FIG. 1. (Continued)

(c) The specificity of active fractions was determined by gel electrophoresis method 2 (Section III,B,2).

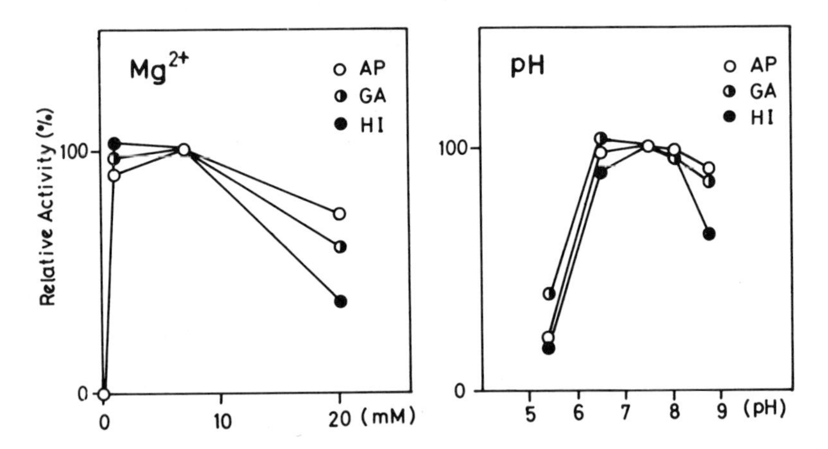

FIG. 2. Effects of Mg ions and pH on enzyme activity. The assay conditions were the same as those used for the transfection assay (Section III,A), except that different concentrations of Mg ions or 50 mM Tris-acetate buffer (with appropriate pH's) were added. After incubation for 20 min at 37°C with 1 unit of enzyme, the remaining infectious DNA was assayed by transfection. Enzyme activity was compared with that exhibited under standard assay conditions.

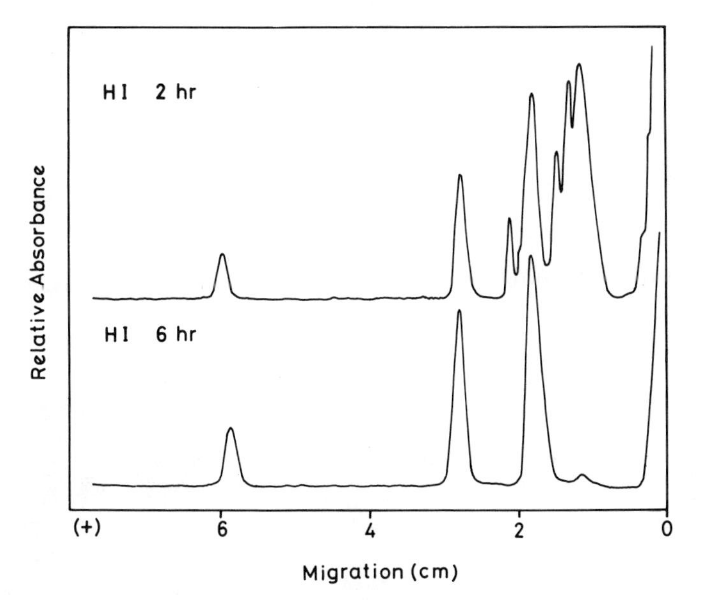

FIG. 3. Gel electrophoresis of Endo H-I hydrolysates of fd RF I DNA. fd RF I DNA (0.3 OD_{260} unit) was incubated with 10 units of Endo H-I for the indicated periods. The hydrolysates were electrophoresed on 3% gel columns. After staining with acridine orange, the color density was traced by a densitometer.

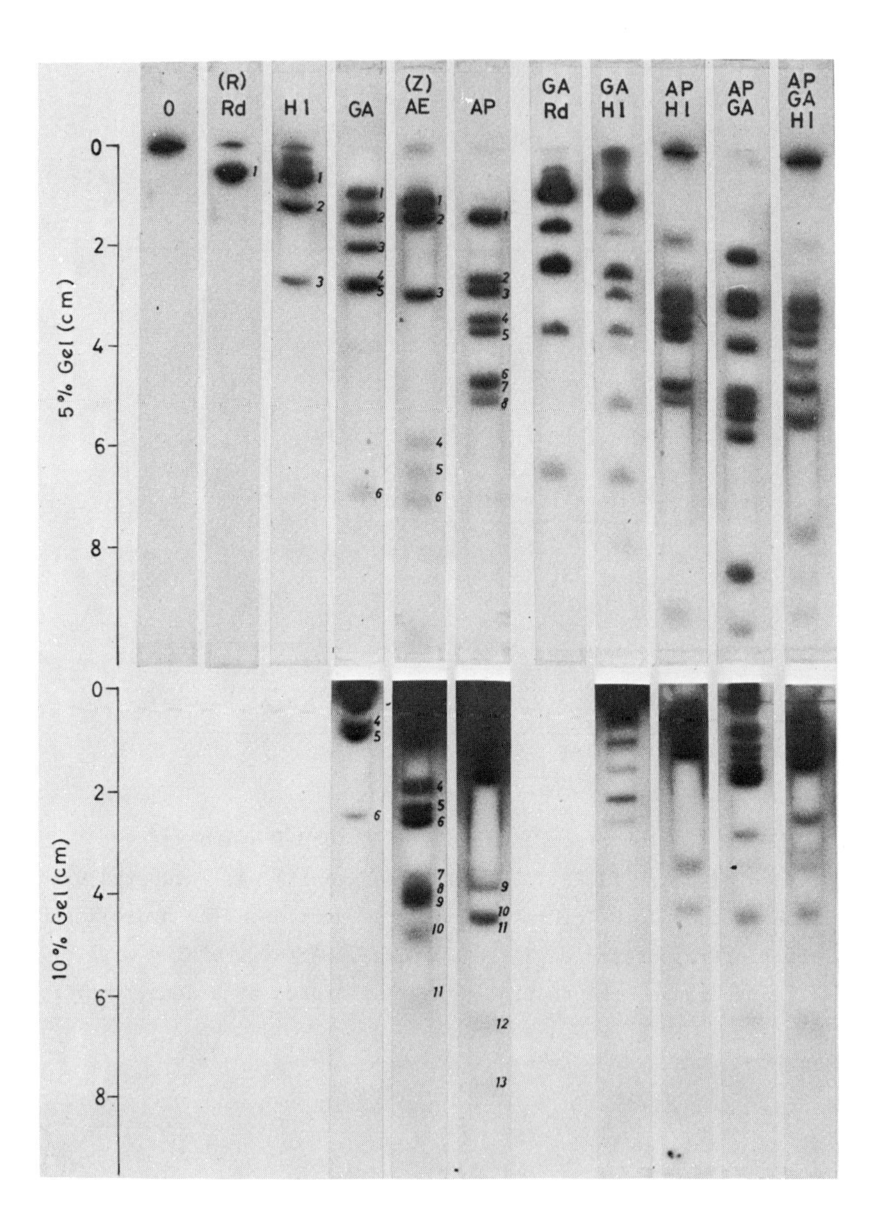
0
(R) Rd
H I
GA
(Z) AE
AP
GA Rd
GA HI
AP HI
AP GA
AP GA HI
5% Gel (cm)
10% Gel (cm)

FIG. 4. Gel electrophoresis of fragments produced from fd RF I DNA by cleavage with five different *Haemophilus* enzymes. fd RF I DNA uniformly labeled with ^{32}P ($\sim 10^5$ cpm/0.03 OD_{260} unit) was hydrolyzed by incubation for 4 hr with more than 20 units of enzymes. The hydrolysates were electrophoresed on 5% and 10% gel columns, and radioautographs were taken as in Section III,B,2. O refers to the original DNA.

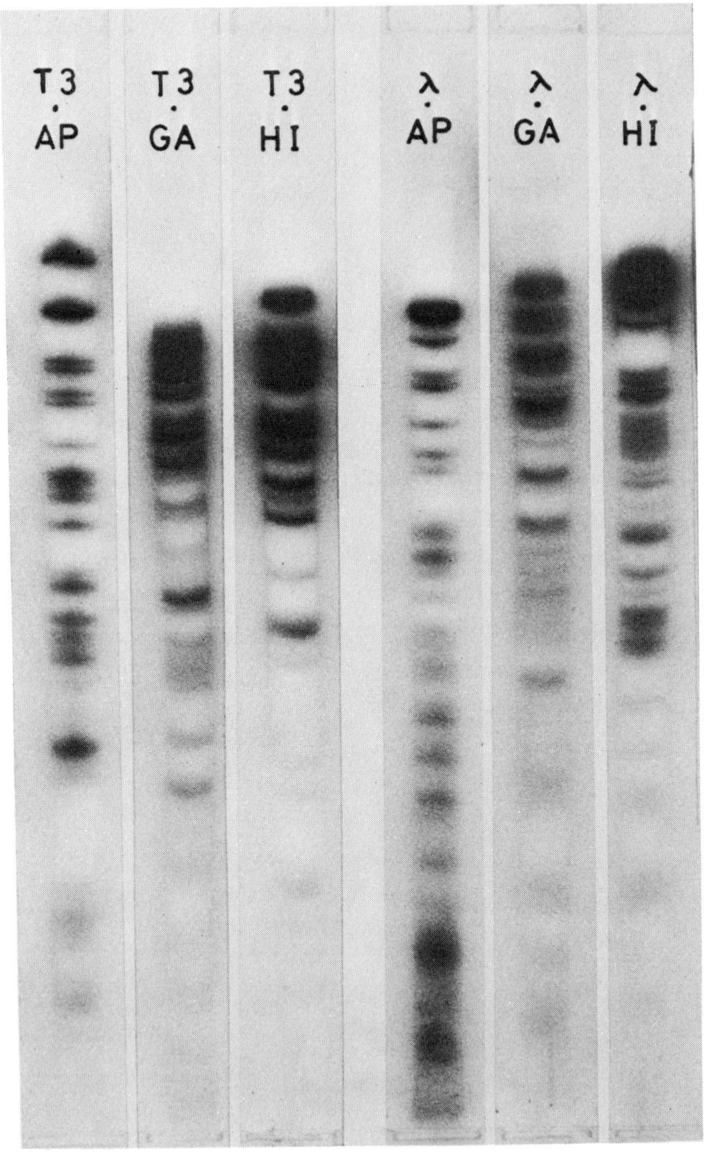

T3·AP T3·GA T3·HI λ·AP λ·GA λ·HI

FIG. 5. Gel electrophoresis of fragments produced from T3 and λ DNA by cleavage with Endo AP, GA, and H-I. T3 and λ DNA uniformly labeled with ^{32}P ($\sim 10^5$ cpm/0.02 OD_{260} unit) were hydrolyzed by incubation with more than 20 units of enzymes for 4 hr. The hydrolysates were electrophoresed on 5% gel columns, and radioautographs were taken as in Section III,B,2.

Chapter 6

DNA SYNTHESIS IN TOLUENE-TREATED CELLS OF *ESCHERICHIA COLI*

Robb E. Moses

Marrs McLean Department of Biochemistry
Baylor College of Medicine
Texas Medical Center
Houston, Texas

I. INTRODUCTION

The bacterial chromosome appears to replicate bi-directionally in a semiconservative manner from a single fixed point in a linear sequential fashion until completely duplicated. It has been postulated that the site for initiation of DNA synthesis is attached to the cell membrane, and newly replicated DNA has been isolated in association with membrane fractions by a variety of procedures. Initiation of DNA synthesis appears to be coupled to cellular division and requires protein synthesis and energy production. It presently appears that RNA synthesis may play a primary role in the initiation of DNA synthesis. These points have been discussed in a number of reviews [1-10].

DNA replication appears to be a complex process involving gene products that may function independently or in aggregate. The discrete processes involved in the replication of the genome have not been defined. The use of synthetic templates and highly purified enzymes of nucleic acid metabolism has not produced satisfactory extensive replicative DNA synthesis.

Difficulty in reassembling the proper components for DNA replication in vitro, as well as ignorance of what these components may be has led to efforts to develop satisfactory in situ or in vitro DNA replicating systems. Such a system must meet the criteria defining replicative synthesis as determined from studies in vivo. In addition it should offer a number of advantages over live cells: (a) It should allow a definition of substrates and the control of the concentration of such substrates; (b) it should bypass the selective permeability of the intact cell, allowing the study of constituents not normally taken up by bacteria; (c) it should allow studies with minimally altered or damaged cellular genomes; (d) it should maintain the required conformation of the replicating complex; (e) it should allow the precise regulation of the rate of DNA synthesis through manipulation of required constituents and temperature; and (f) it should allow isolation of the products of replicative DNA synthesis with greater ease. Several systems meeting some or all of these criteria have been developed for bacteria [11-17].

Toluene treatment of bacterial cells is one method of producing such a system. Toluene treatment kills cells of *E. coli* and makes them permeable to molecules of low molecular weight including the deoxynucleoside 5'-triphosphates, satisfactory precursors for DNA synthesis. Although such cells are no longer viable, they maintain their structure and a number of their physiologic functions, including the ability to synthesize DNA [13,18].

II. PREPARATION OF TOLUENE-TREATED CELLS AND ASSAY FOR DNA SYNTHESIS

A. Toluene Treatment

E. coli cells are prepared [13] by a modification of the method described by Levin et al. [19]. The cells are grown to concentrations of 5 x 10^8 cells/ml and harvested by centrifugation at 10,000 *g* for 15 min at 4°C. The cell pellet is resuspended in 0.05 M KPO4, pH 7.4 to a final volume approximately 10% of that of the harvested culture volume. The cell suspension is made 1% in toluene and gently agitated at 25° for 10 min. Satisfactory agitation may be achieved by shaking the suspension in a test tube on a platform shaker. The temperature during exposure to toluene is not critical; the procedure may be satisfactorily carried out at temperatures ranging from 4° to 37°C. After a 10-min exposure to toluene, the cells are harvested by centrifugation at 10,000 *g* for 15 min at 4°C, the cell pellet is washed with 0.05 M KPO4, pH 7.4, and the pellet is resuspended in the same volume of this buffer as before (a 10- to 15-fold increase in concentration over culture).

The extent of permeability introduced by the toluene treatment may be improved by increasing the surface area during agitation or by preparing an emulsion of toluene and water by sonic irradiation and adding this to the cell suspension [20]. Either of these procedures will decrease the amount of exposure required to achieve satisfactory permeability in the cell. However both of these modifications will also increase the rate of loss of the capacity of the cells for replicative DNA synthesis. For most strains of *E. coli*, a period of exposure in the standard 1% toluene suspension of between 5 and 10 min at 25°C proves satisfactory for optimal replicative DNA synthesis.

After toluene-treatment and resuspension, the cell suspension is divided into aliquots and frozen at -80°C. *E. coli* cell suspensions prepared and stored in this manner are stable for periods of greater than one year. This permits the use of a well-defined preparation of cells for a number of studies. For most strains of *E. coli* that we have used, freezing and thawing does not result in a rapid loss of replicative DNA synthesis in the cell preparation if care is taken to maintain the thawed suspension at 4°C. It is important that as much of the toluene as possible be removed during the washing, since remaining toluene decreases the stability of the preparation. Toluene-treated cells may be assayed directly without washing free of toluene, but such preparations are much less stable for storage, losing their DNA synthesizing activity at the rate of approximately 50% in 30 min at 4°C.

Due to some variability in the sensitivity of various strains of *E. coli* to toluene treatment, it is best when preparing a large batch of toluene-treated cells to do a time course of toluene treatment versus DNA synthesis measured in vitro in a small trial before carrying out the large preparation. Treatment of the cells with toluene will usually lead to a decrease in replicative DNA synthesis after 10 to 12 min of exposure to toluene under the conditions described; a few strains will show a maximal uncovering of replicative DNA syntheis in as little as 2 min. Still other strains, such as strain BT1026, a $polA^-$ $polC_{ts}$ mutant (DNA polymerase I deficient, DNA polymerase III temperature sensitive), will not be stable to washing after toluene treatment and must be assayed directly after treatment without washing [21].

B. Assay of DNA Synthesis

The standard reaction mixture (0.3 ml) contains 70 mM KPO_4, pH 7.4; 13 mM $MgCl_2$; 1.3 mM ATP; 33 μM dCTP, dGTP, dATP, $[^3H]$dTTP; and 1 to 5 x 10^8 toluene treated cells. The radioactive label, satisfactorily used on any one of the deoxynucleoside 5'-triphosphates, has a specific activity of from 1 to 10 x 10^4 cpm/nmole. Reaction mixtures may be warmed before the addition of toluene-treated cells or may be completed at 4°C and brought to the required temperature in the water bath. Standard incubations are for 30 min at 37°C. Reactions are terminated by the addition of 3 ml of cold 10% trichloroacetic acid-0.1 M sodium pyrophosphate. After addition of acid, the mixes are allowed to stand for 5 min at 4°C; the precipitated material is then collected by passing over a Whatman GF/C (2.4 cm) disk under the force of aspirator vacuum. The disk is washed twice with 3 ml portions of 10% trichloroacetic acid-0.1 M sodium pyrophosphate and 10 ml of cold 0.01 M HCl. The disks are dried and collected, and radioactivity is determined in a toluene-based scintillation fluid containing 4 g of 2,5-diphenyloxazole and 0.05 g of 1,4-bis 2-(5-phenyloxazoyl) benzene per liter. Typical assays may demonstrate the incorporation of several hundred picomoles of one labeled deoxynucleoside 5'-triphosphate.

C. DNA Synthesis in Triton X-100

In order to increase the permeability of the toluene-treated cells to macromolecules, replicative DNA synthesis may be measured in reaction mixtures that contain the nonionic detergent Triton X-100 [22]. This detergent (p-isocetylpolyoxyethylenephenol) appears to have no detrimental effect upon replicative synthesis [23]. Care must be taken to ensure adequate mixing in reaction mixtures containing the detergent, or two phases will persist. A level of 1% detergent is routinely used. Although we do not routinely add 2-mercaptoethanol to our reaction mixtures, when Triton X-100 is present in the reaction mixture, 10 mM 2-mercapto-ethanol is included. The sulfhydryl reagent in the presence of detergent improves the linearity of the DNA synthesis reaction.

D. Requirements for Replicative DNA Synthesis

All four deoxynucleoside 5'-triphosphates are required for replicative DNA synthesis (Table 1). In addition Mg^{2+}, K^{+}, and ATP are required for optimal DNA synthesis. The requirement for ATP is a striking feature, synthesis showing 10- to 20-fold stimulation in the presence of this compound. This requirement is not understood although it has been extensively investigated [11-14,16,17,24]. The ATP stimulation is not due to the *rec B,C* gene product, because such stimulation for replicative DNA synthesis is observed in mutants deficient in these gene products [13,25]. The requirement for ATP is not highly specific since other nucleoside triphosphates can partially substitute for ATP [13,14,24]. However, analogs of nucleoside triphosphates do not appear to substitute for this requirement [14].

TABLE 1

Requirements for Replicative DNA Synthesis in Toluene-Treated *E. coli*[a]

Condition	Relative activity, %
Control	100
-ATP	5
-4 dNTP, +4 dNDP	95
$-Mg^{2+}$	3
+3 mM NEM	3
Toluene treatment	2
-3 dNTP	<1
+ Naladixic acid	35
+ Sodium azide	102
+ Sodium cyanide	95
+ Rifampicin	92

[a]Reaction conditions were as described: (0.3 ml) 66 mM potassium phosphate buffer, pH 7.4; 13 mM $MgCl_2$; 1.3 mM ATP; 33 μM dCTP, dGTP, dATP, [^{3}H]dTTP; and cells (*E. coli* P3478). Incubations were carried out for 30 min at 37°C. 100% activity is 120 pmoles of [^{3}H]dTTP. NEM is N-ethylmaleimide. Naladixic acid was present at 90 μg/ml; sodium azide was present at 1 mM; sodium cyanide was present at 0.2 mM; rifampicin was present at 100 μg/ml.

The deoxynucleoside 5'-diphosphates substitute well for the triphosphates (Table 1), while the deoxynucleoside monophosphates substitute partially in some instances. Replicative DNA synthesis in toluene-treated cells of *E. coli* is sensitive to inhibition by sulfhydryl blocking agents such as N-ethylmaleimide or p-chloromecuribenzoate. There is a requirement for K^+ and the substitution of sodium phosphate for potassium phosphate buffer is a common cause for inexplicably poor replicative DNA synthesis in toluene-treated cells.

When first using a large preparation of cells in this in vitro system, it is advantageous to do trial reactions varying the concentrations of Mg^{2+} and ATP. The ratio of these reactants as well as the absolute levels will affect the rate and the extent of the reaction. We routinely maintain the ratio of Mg^{2+}/ATP at 10:1 during such trials. The requirement for ATP for optimal replicative DNA synthesis will vary from cell preparation to cell preparation but is usually in the range of 1 to 2 mM. It seems likely that an inhibition at higher concentration results from competition for triphosphate-binding sites in the replicative system.

III. MODES OF SYNTHESIS

A. Replicative DNA Synthesis

DNA synthesis occurring under the reaction conditions given above is predominantly replicative in nature. By a number of criteria it mimics DNA synthesis occurring during growth of intact *E. coli* cells. It produces a high-molecular-weight product, it is semiconservative in nature, it is thermolabile in *dna*$_{ts}$ mutants of the "fast-stop" variety, it occurs at a rate approaching in vivo DNA synthesis, it is sensitive to naladixic acid, a 10S species of DNA can be observed, and synthesis continues at the growing point operative just prior to toluene treatment [13,18,26]. In a modification of this system applicable to *Bacillus subtilis*, active transforming DNA has been synthesized with the same transforming activity as DNA made in vivo [27].

In *polA1* strains of *E. coli* [28], replicative DNA synthesis is the predominant mode of DNA synthesis observed in the system described. In the absence of ATP or in the presence of N-ethylmaleimide, only 2 to 5% of the optimal rate of DNA synthesis is observed in this system. Ten to 11S fragments of DNA (Okazaki fragments) have been observed in toluene-treated cells during replicative DNA synthesis [24,29-31]. The conversion of these fragments of DNA to high-molecular-weight DNA follows a pattern similar to that observed in vivo, but the toluene-treated cell system allows blockage of the conversion by the addition of nicotinamide mononucleotide (NMN) to discharge ligase-AMP. These observations illustrate an advantage of the in vitro system, i.e., metabolites may be added at much higher concentrations than they occur in vivo to study the effects upon the metabolism of the DNA. The data in this instance suggest that ligase plays a prominent role in the conversion of Okazaki pieces of DNA to high-molecular-weight DNA.

Low-molecular-weight ribonucleotide polymers have been observed in toluene-treated cells, and their formation and linkage to single-stranded DNA suggests a role as primer for DNA synthesis [31]. These observations are comparable with those in a number of other systems, suggesting that ribonucleotide polymers are responsible for microinitiation of DNA synthesis.

Host-range restriction may be demonstrated in vitro in toluene-treated cells [32]. If hydroxymethyldeoxycytidine 5'-triphosphate (dHMCTP) is substituted for dCTP in the reaction mixture, replicative DNA synthesis is lowered to less than 5% of normal levels in wild-type cells. However, in strains permissive for the growth of nonglucosylated T-even bacteriophage, normal levels of DNA synthesis are observed and dHMCTP is incorporated into the product.

Perhaps the data that best support the conclusion that DNA synthesis in the toluene-treated cell mimics in vivo replication is the behavior of DNA synthesis in this system in cells of E. coli mutants that are temperature sensitive for DNA replication in vivo. In mutants of the "fast-stop" variety (dnaB and dnaG), replicative DNA synthesis ceases rapidly in the in vitro system [13,22,26]. The kinetics of arrest of DNA synthesis in the in vitro system duplicate those observed in vivo. In DNA temperature-sensitive mutants of the "slow-stop" variety (dnaA), replicative DNA synthesis in the toluene-treated cell system follows a pattern similar to wild type upon a shift to the restrictive temperature (42°C). Mutants of this class have a defect in initiation of new rounds of chromosomal copying in vivo at the restrictive temperature. Since there is no indication of such initiation occurring in the in vitro system (it would require protein synthesis), it is not surprising that a defect is not observed in the toluene-treated cell system.

Toluene-treated cells provide a useful tool for the investigation of the effect of drugs on DNA synthesis. For example, naladixic acid inhibits replicative DNA synthesis but rifampicin does not [24,29,34]. 1-β-D-Arabinofuranosylcytosine 5'-trisphosphate appears to inhibit DNA replication in the toluene-treated cell system [33].

The increase in permeability in cells incubated in the presence of Triton permits modification of synthesis through the addition of macromolecules. DNA polymerase I activity in toluene-treated cells may be inhibited to a few percent or less of original levels through the addition of antibody to purified DNA polymerase I [23]. This antibody does not inhibit replicative DNA synthesis. Although replicative DNA synthesis may be measured in toluene-treated cells that contain DNA polymerase I as well as those that do not, modification of the system through the addition of Triton and antibody allows the in vitro construction of a DNA polymerase I-deficient strain. This promises to be of advantage in instances when it is difficult to construct a double mutant with a DNA polymerase I deficiency (polA1) due to inviability of the double mutant, for example, recA$^-$ polA1. The antibody has been shown to inhibit both the polymerizing activity and the nuclease activities of DNA polymerase I in purified enzyme systems in vitro [5]. This may be of advantage because recent results [35] have indicated that polA1 mutants of E. coli contain nearly normal levels of 5'-3' exonucleolytic activity. The Triton-antibody modification of the toluene-treated cell system should allow an investigation of whether DNA polymerase I nucleolytic activity is required for the maintenance of replicative DNA synthesis after exposure to certain agents.

B. Repair Synthesis

In the toluene-treated cell system it is possible to elicit a second type of DNA synthesis distinct from replicative DNA synthesis. This synthesis is dependent upon DNA polymerase I and either endogenous or exogenous nuclease activity. It is not semiconservative in nature, it is not ATP stimulated, and it is not sensitive to sulfhydryl-blocking agents. Such synthesis is most reproducibly elicited through the use of an endonuclease I-deficient strain of E. coli and the addition of pancreatic DNase [13]. The addition of nuclease to toluene-treated cells results in the destruction or disappearance of replicative DNA synthesis and replacement by a nonconservative synthesis that is dependent upon DNA polymerase I activity. For these reasons, this synthesis has been termed "repair" synthesis in light of the probability that it is DNA synthesis occurring at nicks and uneven ends introduced into the genome of the cell. Synthesis by DNA polymerase I at such loci would be representative of the enzymatic repair process in purified enzyme systems [3,5]. The level of pancreatic nuclease used to obtain satisfactory levels of repair synthesis ranges from 0.1 to 0.5 μg/ml of reaction mixture. In contrast to replicative synthesis, the nuclease-dependent DNA synthesis observed in toluene-treated cells is not thermolabile in DNA temperature-sensitive mutants of the "fast-stop" variety. This synthesis probably is similar to the repair synthesis observed in other permeable cell systems [36].

Toluene-treated cells of *E. coli* have also been used to investigate DNA synthesis after ultraviolet irradiation. Qualitatively similar results are obtained whether the cells are irradiated before or after toluene treatment [37,38]. Ultraviolet irradiation produces marked inhibition of replicative DNA synthesis in toluene-treated *E. coli* cells, but such irradiation elicits the appearance of a novel type of DNA synthesis. Ultraviolet irradiation-stimulated DNA synthesis in toluene-treated cells is not dependent upon DNA polymerase I activity since the synthesis is observed in DNA polymerase I-deficient mutants. Such synthesis is nonconservative in nature and is independent of replication as shown by its occurrence at the restrictive temperature in DNA temperature-sensitive mutants of the "fast-stop" variety. Post-irradiation synthesis occurs normally in toluene-treated cells of strains [20] that are deficient in DNA polymerase II.

ACKNOWLEDGMENTS

Parts of this work carried out in the author's laboratory were supported by National Institutes of Health, United States Public Health Service Grant No. GM 19122, Grant No. VC-97 from the American Cancer Society, Inc., and Robert A. Welch Foundation Grant No. Q-543. R.E.M. is the recipient of a Public Health Service Research Career Program Award, No. GM 70314.

REFERENCES

1. F. Jacob, S. Brenner, and F. Cuzin, Cold Spring Harbor Symp. Quant. Biol., 28, 329 (1963).

2. A. Ryter, Bacteriol. Rev., 32, 39 (1968).

3. C. C. Richardson, Ann. Rev. Biochem., 38, 795 (1969).

4. K. G. Lark, Ann. Rev. Biochem., 38, 569 (1969).

5. M. Goulian, Ann. Rev. Biochem., 40, 855 (1971).

6. J. D. Gross, Current Topics Microbiol. Immunol., 57, 39 (1972).

7. F. Bonhoeffer and W. Messer, Ann. Rev. Genetics, 3, 233 (1969).

8. N. Sueoka, Molecular Genetics, Vol. 2 (J. H. Taylor, ed.), Academic, New York, 1967, p. 1.

9. J. Wechsler and J. D. Gross, Molec. Gen. Genetics, 113, 273 (1971).

10. M. L. Pato, Ann. Rev. Microbiology, 26, 347 (1972).

11. D. W. Smith, H. E. Schaller, and F. J. Bonhoeffer, Nature, 226, 711 (1970).

12. R. Knippers and W. Stratling, Nature, 226, 713 (1970).

13. R. E. Moses and C. C. Richardson, Proc. Natl. Acad. Sci. U.S., 67, 674 (1970).

14. H.-P. Vosberg and H. Hoffmann-Berling, J. Mol. Biol., 58, 739 (1971).

15. A. T. Ganesan, Proc. Natl. Acad. Sci. U.S., 68, 1296 (1971).

16. H. Schaller, B. Otto, U. Nusslein, J. Huf, R. Hermann, and F. Bonhoeffer, J. Mol. Biol., 63, 183 (1972).

17. R. B. Wickner and J. Hurwitz, Biochem. Biophys. Res. Commun., 47, 202 (1972).

18. J. Mordoh, Y. Hirota, and F. Jacob, Proc. Natl. Acad. Sci. U.S., 67, 773 (1970).

19. D. H. Levin, M. N. Thang, and M. Grunberg-Manago, Biochim. Biophys. Acta, 76, 558 (1963).

20. J. L. Campbell, L. Soll, and C. C. Richardson, Proc. Natl. Acad. Sci. U.S., 69, 2090 (1972).

21. R. E. Moses, unpublished observation.

22. Triton X-100 was obtained from Rohm and Haas, Philadelphia, Pennsylvania.

23. R. E. Moses, J. Biol. Chem., 247, 6031 (1972).

24. D. Pisetsky, I. Berkower, R. Wickner, and J. Hurwitz, J. Mol. Biol., 71, 557 (1972).

25. G. W. Bazill, R. Hall, and J. Gross, Nature New Biol., 233, 283 (1971).

26. R. Burger, Proc. Natl. Acad. Sci. U.S., 68, 2124 (1971).

27. T. Matsushita, K. P. White, and N. Sueoka, Nature New Biol., 232, 111 (1971).

28. P. DeLucia and J. Cairns, Nature, 224, 1164 (1969).

29. R. E. Moses, J. L. Campbell, R. A. Fleischman, and C. C. Richardson, in Miami Winter Symposia, Vol. 2 (D. W. Ribbons, J. F. Woessner, and J. Shultz, eds.), North-Holland, Amsterdam, 1971, p. 48.

30. J. L. Campbell, R. A. Fleischman, and C. C. Richardson, Federation Proc., 30, 1313 (1971).

31. A. Sugino and R. Okazaki, Proc. Natl. Acad. Sci. U.S., 70, 88 (1973).

32. R. A. Fleischman and C. C. Richardson, Proc. Natl. Acad. Sci. U.S., 68, 2527 (1971).

33. G. V. R. Reddy, M. Goulian, and S. S. Hendler, Nature New Biol., 234, 286 (1971).

34. A. M. Pedrini, D. Geroldi, A. Siccardi, and A. Falaschi, Europ. J. Biochem., 25, 359 (1972).

35. I. R. Lehman and J. R. Chien, Federation Proc., 32, 1282 (1973).

36. G. Buttin and A. Kornberg, J. Biol. Chem., 241, 5419 (1966).

37. W. E. Masker and P. C. Hanawalt, Proc. Natl. Acad. Sci. U.S., 70, 129 (1973).

38. D. Bowersock and R. E. Moses, J. Biol. Chem., 248, 7449 (1973).

Chapter 7

ETHER-TREATED CELLS

Hans Peter Vosberg* and H. Hoffmann-Berling

Max-Planck-Institut für Medizinische Forschung
Abteilung Molekulare Biologie
Heidelberg, West Germany

*Present address: California Institute of Technology, Division of Biology, Pasadena, California.

I. INTRODUCTION

When *Escherichia coli* cells are brought into contact with diethylether, lipid is extracted, most likely from the bacterial envelopes, free nucleotides leak out from the cells, and synthetic processes cease. Studies on these ether-permeabilized cells have shown that they resume DNA synthesis upon addition of deoxynucleoside triphosphates (dNTP) and that they are able to perform extended semiconservative as well as repair-type synthesis [1-3].

Ether-treated cells offer advantages in pulse-labeling studies on DNA structure: absorbed radioactivity fails to be diluted into a pool of endogenous DNA precursors and synthesis during the labeling procedure can be adjusted to a conveniently low rate by applying dNTPs at low concentrations. Ether-treated cells have further been used for studying the effects of nucleotide analogs on DNA synthesis [4].

Semi-conservative synthesis, an ATP-dependent process, is the main, and possibly the sole, type of synthesis under the assay conditions specified below and can be studied equally well in mutant cells deficient for DNA polymerase I and in the corresponding wild-type cells. Repair-type synthesis, an ATP-independent process, occurs in cells that contain a normal level of DNA polymerase I as a response to nucleolytic cleavage in DNA.

Two kinds of templates are used in the procedures to be described for studying semiconservative synthesis in ether-treated cells: *E. coli* DNA and the double-stranded replicative form (RF) DNA of bacteriophage ØX174. This small, circular DNA replicates semiconservatively (RF to RF) using host enzymes in conjunction with viral gene A product [5]. In addition, techniques are described for studying the conversion of the single-stranded DNA (SS) of ØX174 to RF by synthesis of a complementary strand. This early step of viral infection is catalyzed entirely by host enzymes and shares with replication synthesis of *E. coli* DNA the requirement for several *dna* gene products (demonstrated in cell extracts) and for ATP (demonstrated in cell extracts and ether-treated cells) [6-8].

II. CHOICE OF BACTERIAL STRAINS

Ether treatment leads to an activation of the bacterial endonuclease I, which then cleaves cellular DNA. In order to avoid this unwanted effect, endonuclease I deficient (*endI*) bacterial strains are used. Furthermore, in some of the procedures to be described mitomycin C is applied to cells in order to suppress, preferentially, the synthesis of host DNA after infection with ØX [9]. This technique requires ultraviolet reactivation-deficient (*uvr*) cells.

The ØX-sensitive K12/C hybrid strain H512 (*endI uvrA arg*) is well suited for both these reasons [2]. A DNA polymerase I-deficient (*polA*) strain suited for some of the procedures is the ØX-sensitive K12/C hybrid H560 (*endI* $polA_1$ *tsx str*) [2]. This strain is uvr^+.

III. PREPARATION OF ETHER-TREATED CELLS

The procedures vary according to the template to be studied, and in some steps they are modified versions of published techniques.

A. Semiconservative Synthesis of E. coli DNA

Cells are grown with aeration at 37°C in 800 ml of a minimal salt medium [10] containing 0.05 M glycerol as an energy source and enriched with 0.4% casamino acids (Difco) ("growth medium"). At a density of 3 x 10^8 cells/ml the culture is transferred to a cold room and the cells are collected by sedimentation at 3500 g for 5 min. Cell pellets are suspended by pipeting in 15 ml KTM [KCl 80 mM, Tris-HCl 40 mM, magnesium acetate 7 mM, spermidine-3 HCl 0.4 mM, EGTA 2 mM (pH 7.6)] containing 0.5 M sucrose and transferred to a glass-stoppered separatory funnel. Ether purissimum pro narcosi (Hoechst, Darmstadt) 20 ml, kept at -18°C until use, is added and the two phases are mixed by gentle agitation for 30 sec. An additional 10 ml of KTM-0.5 M sucrose solution is added to promote phase separation. The lower watery phage is then layered over a 10 ml cushion of 0.75 M sucrose in KTM and the layers are centrifuged in a Sorvall instrument for 8 min at 3500 g. The pelleted cells are suspended by gentle pipeting in 4 ml KTM-0.5 M sucrose and this suspension, which contains about 5 x 10^{10} cells/ml, is frozen in small portions and thawed only once. Storage at -18°C for several weeks does not affect the capacity for DNA synthesis [2].

EGTA (ethyleneglycol-bis(2-aminoethylether) tetraacetic acid; Serva, Heidelberg), which complexes Ca^{2+} but not Mg^{2+}, is present to prevent heavy metal ion effects; sucrose and spermidine are also present to guard against bacterial lysis. Ether contains 5 mg/liter of 2,6-tert-butyl-4-methyl phenol as an antioxidant.

B. Semiconservative Synthesis of ØX RF-DNA

Two steps are introduced between collection of the cells and their resuspension in the KTM medium. (a) Pelleted cells are suspended in 40 ml starvation buffer (KCl 67 mM; NaCl 17 mM; Tris-HCl 10 mM; $MgSO_4$ 0.4 mM; $CaCl_2$ 1 mM; pH 8.1) [11] at 37°C followed by short aeration (5 to 15 min). Mitomycin C, 50 μg/ml (Kyowa, Tokyo), from a tenfold concentrated stock, is then added to inhibit preferentially the synthesis of host DNA, and incubation is continued for 12 min without aeration in the dark. Mitomycin treatment is omitted with uvr^+ strains. (b) The cells are sedimented and resuspended in 40 ml of fresh starvation buffer, and ØX phage (3 infectious units per cell) is added in 0.1 vol starvation buffer and allowed to adsorb at 37°C without aeration for 3 min. The infected cells are then diluted into 350 ml of warm aerated growth medium to initiate rapid viral development. Five minutes later, when RF replication has started, the culture is poured onto 0.3 vol of frozen, crushed growth medium and the cells are collected for ether treatment [2,12].

C. Conversion of ØX SS to RF

After centrifugation of cells during the standard preparation procedure, they are suspended in 40 ml of warm growth medium containing caffeine, 15 mg/ml, which blocks DNA, RNA, and protein synthesis and, thus, replication of virus. Phage is added 3 min later in 0.1 vol of growth medium at a multiplicity of five. After standing for 2 min, the mixture is diluted into 350 ml warm growth medium containing caffeine, 15 mg/ml, and aerated for 5 min to allow the infectious virus to go into eclipse. Following sedimentation and suspension in 15 ml cold KTM containing 0.5 M sucrose and caffeine, 15 mg/ml, the cells are treated with ether and collected by sedimentation through an underlayer of sucrose in KTM (see above) [8].

Immediately before using caffeine is dissolved in the KTM-0.5 M sucrose solution at 100°C followed by cooling, since the caffeine precipitates from this solution on standing.

IV. ASSAY FOR DNA SYNTHESIS

The standard assay mixture (35°C) is prepared in KTM with 0.5 M sucrose; it contains 5 x 10^9 ether-treated cells/ml and 20 μM each of four dNTPs, one of which is usually included as a ^{3}H- or α^{32}P-labeled compound (about 10 μCi/ml). ATP 1 mM is also added as a cofactor for semiconservative synthesis and to stabilize dNTP. DNA synthesis is measured as acid-precipitable radioactivity after pipeting 0.1 ml samples into 2 ml cold 0.1 M NaCl and adding 1 vol 10% trichloroacetic acid (TCA) that contains 2 mg RNA hydrolysate/ml (prepared as a tenfold stock by heating commercial RNA in 5% TCA for 10 min at 90°C). The precipitate is washed five times with cold 5% TCA-RNA hydrolysate (1 mg/ml) on Schleicher and Schuell B 185 membrane filters or Whatman GF/C glass filters, dried, and counted.

The counting efficiency of ^{3}H in a toluene-based scintillator (Ciba butyl-PBD) is 15 to 18% on filters loaded with 5 x 10^8 cells.

V. EXTRACTION OF DNA FOR ANALYSIS

Synthesis is stopped with 4 vol of a cold solution containing 50 mM Tris-HCl and 5 mM EDTA (pH 8.1). The cells are sedimented and resuspended in this medium to a density not higher than 10^{10} cells/ml and lysed with lysozyme, 0.2 mg/ml, by incubating first for 10 min in ice and then for 30 min at 37°C. DNA is extracted by incubating overnight with predigested [13]

Pronase P (100 μg/ml; Kaken, Tokyo) in the presence of 1% sarcosyl NL97 (Serva, Heidelberg) at 37°C or by phenol treatment.

In studies on parental-labeled ØX DNA, it is sometimes desirable to elute label of adsorbed, noneclipsed phage from the cells before subjecting them to lysis. Elution of some label is achieved by washing the cells three times with a freshly prepared solution of 0.05 M sodium tetraborate containing 0.005 M EDTA and 0.5 M sucrose [14]. Furthermore, only phenol treatment completely frees intermediates of SS-to-RF conversion from protein. Phenol treatment is carried out after overnight digestion with Sarkosyl-pronase. This procedure was found to reduce the sensitivity of the intermediate complexes to shear.

VI. GENERAL COMMENTS

Cells made permeable with toluene to nucleotides [15] and ether-permeabilized cells are probably equivalent with regard to capacity for DNA synthesis. Ether treatment has the advantage of interrupting in vivo synthesis within a few seconds. Ether-treated cells lose little endogenous protein, suggesting that few cells are permeable to macromolecules [2].

A. Semiconservative synthesis of E. coli DNA

Studies on the DNA product synthesized in ether-treated endI cells from exogenous deoxynucleotides have led to the following conclusions. Synthesis starts from sites of in vivo replication on the chromosome and proceeds in a discontinuous fashion. The fragments of new DNA that appear under the assay conditions specified above have approximately the same mean length as Okazaki pieces. They transfer incorporated label to longer units by a reaction that is inhibited by nicotinamide mononucleotide, suggesting that the fragments are joined together and that joining requires the activity of DNA ligase. The final product is a long continuous chain of new DNA base paired to a parental chain [3,4].

The maximal rate of synthesis in these cells is about 500 nucleotides incorporated per cell per second at 35°C as compared with about 2000 nucleotides per replication fork per second in intact cells. Depending on the bacterial strain, synthesis continues for 15 to 60 min [2].

Specific inhibition of this synthesis is achieved by: (a) Starvation [16] of amino acid-requiring cells for 100 min before ether treatment [17]; initiated rounds of chromosomal replication are completed under these conditions, while no reinitiation occurs [18]. (b) Another possibility is to shift to the nonpermissive temperature (40°C) of permeabilized cells that are mutants defective in dnaB, dnaE (the gene for DNA polymerase III), or dnaG [19]. These genes provide functions for the continuation of initiated chromosomal replication [20]. (c) Finally, mitomycin C inhibits as described above.

By the addition of pancreatic DNase (1 μg/ml) to the assay mixture for DNA synthesis, we can provoke starved or mitomycin-treated permeable $polA^+$ cells to perform repair-type synthesis. The resulting 20- to 100-fold stimulation of synthesis is not observed with polA cells [1,2,16]. The low rate of synthesis measured without this activating treatment in the permeable $polA^+$ cells ($<10\%$ of the synthesis measured in cells collected from an exponential culture) is the main reason for the belief that repair-type synthesis is negligible under normal assay conditions in ether-treated $polA^+$ cells, which are endI.

The genetic evidence quoted above suggests that the function of DNA polymerase III is required for semiconservative synthesis in the permeable cells. Whereas this enzyme and its modified form, DNA polymerase III*, are 90% inhibited by 0.13 M KCl [21,22], semiconservative synthesis has an optimal salt concentration of between 0.08 and 0.18 M KCl.

B. Semiconservative Synthesis of ØX RF DNA

This synthesis accounts for 80 to 90% of the total DNA synthesis in mitomycin-treated cells, and about 50% in mitomycin-free cells. Synthesis occurs equally well in polA as in $polA^+$ cells, and it is inhibited by a nonsuppressed mutation in the viral gene A. New RF DNA accumulates as a population of half-synthetic molecules (one new, one conserved strand) containing new viral and new complementary strands in approximately equal numbers. The filling of gaps, a late step in the replication of RF, is detectable. These results are consistent with the view that RF can undergo a complete round of replication in ether-treated cells. The number of RFs that can replicate in a cell has been estimated to approach three. The final RF product is a mixture of relaxed and supercoiled molecules [12, 23,24].

C. SS To RF Conversion

The RF product of this process consists of completely new complementary strands base paired to infecting strands and accumulates as a mixture of relaxed molecules and supercoils. The number of the RFs made is similar to that of the input viable phage. In the nascent state the complementary strand material is discontinuous and, as recently found, contains gaps, suggesting multiple initiations on a given template strand [8,25].

REFERENCES

1. H. Hoffmann-Berling, in Wissenschaftliche Konferenz der Gesellschaft Deutscher Naturforscher und Arzte, Springer, Berlin-Heidelberg-New York, 1968, p. 263.

2. H. P. Vosberg and H. Hoffmann-Berling, J. Mol. Biol., 58, 739 (1971).

3. K. Geider and H. Hoffmann-Berling, Eur. J. Biochem., 21, 374 (1971).

4. K. Geider, Eur. J. Biochem., 27, 554 (1972).

5. R. L. Sinsheimer, in Progress in Nucleic Acid Research and Molecular Biology, Vol. 8, Academic Press, New York, 1968, p. 141.

6. R. Schekman, W. Wickner, O. Westergaard, D. Brutlag, K. Geider, L. L. Bertsch, and A. Kornberg, Proc. Natl. Acad. Sci. U.S., 69, 2691 (1972).

7. R. B. Wickner, M. Wright, S. Wickner, and J. Hurwitz, Proc. Natl. Acad. Sci. U.S., 69, 3233 (1972).

8. U. Hess, H. Durwald, and H. Hoffmann-Berling, J. Mol. Biol., 73, 407 (1973).

9. W. H. Lindqvist and R. L. Sinsheimer, J. Mol. Biol., 30, 69 (1967).

10. H. Hoffmann-Berling, H. Durwald, and I. Beulke, Z. Naturforsch, 19b, 593 (1964)

11. D. T. Denhardt and R. L. Sinsheimer, J. Mol. Biol., 12, 641 (1965).

12. H. Durwald and H. Hoffmann-Berling, J. Mol. Biol., 58, 755 (1971).

13. E. T. Young and R. L. Sinsheimer, J. Mol. Biol., 30, 165 (1967).

14. J. E. Newbold and R. L. Sinsheimer, J. Mol. Biol., 49, 49 (1970).

15. R. C. Moses and C. C. Richardson, Proc. Natl. Acad. Sci. U.S., 67, 674 (1970).

16. P. Rogers, private communication, 1973.

17. R. P. Novick and W. K. Maas, J. Bacteriol., 81, 236 (1961).

18. F. Bonhoeffer and W. Messer, Ann. Rev. Genetics, 3, 233 (1969).

19. H. Durwald, unpublished results.

20. J. A. Wechsler and J. D. Gross, Mol. Gen. Genetics, 113, 273 (1971).

21. T. Kornberg and M. L. Gefter, J. Biol. Chem., 247, 5369 (1972).

22. W. Wickner, R. Schekman, K. Geider, and A. Kornberg, Proc. Natl. Acad. Sci. U.S., 70, 1764, 1973.

23. H. Muller-Wecker, K. Geider, and H. Hoffmann-Berling, J. Mol. Biol., 69, 319 (1972).

24. K. Geider, H. Lechner, and H. Hoffmann-Berling, J. Mol. Biol., 69, 333 (1972).

25. H. P. Vosberg and U. Hess, unpublished results.

Chapter 8

DNA REPLICATION IN THE CELLOPHANE MEMBRANE SYSTEM

Volker Nüsslein and Albrecht Klein*

Max-Planck-Institut für Virusforschung
Tübingen, West Germany

---

*Present address: Lehrstuhl für Mikrobiologie, Universität Heidelberg, Germany.

I. INTRODUCTION

The in vitro system [1] to be described consists of a layer of bacterial cells lysed on a cellophane membrane that is permeable to small molecules. DNA synthesis by the system is performed on a drop of buffer and salt solution containing the necessary precursors.

The system is open to the addition of macromolecules to the lysate on the cellophane membrane. Diffusion of small molecules through the membrane into the lysate is fast because of the thin layer of the lysate obtained. An essential feature of the lysis method used is that the concentration of the macromolecular cell components is kept close to that found in vivo. The DNA remains unbroken and thus repair synthesis in vitro is kept to a minimum. This chapter describes the preparation and use of the system for the in vitro study of DNA replication.

The system is also capable of RNA synthesis, which, however, is not discussed here.

II. MATERIALS

A. Bacterial and Phage Strains

1. BT 1000 (referred to as wild type), *thy* A, *end* A, *pol* A1, *str* A, a thymine-requiring derivative of strain H 560.

2. BT 1026 (*dna* E) and BT 1047 (*dna* B), temperature sensitive derivatives of strain BT 1000 [2].

3. BT 308 thy A, thr, leu, thi, dna G 308 [3].

4. NY 73 thy A, pol A1, str A, thi, leu, met E, dna G3 [3], a gift from Dr. J. A. Wechsler, New York.

5. H 502 thy A, end A.

6. λCI; λCI 0125; λCI P80 [4].

B. Cellophane Membrane Disks

Circular disks, 22 μm thick, cut out of Kalle Einmach Cellophan (Kalle AG, Wiesbaden Biebrich, Germany) were used (Fig. 1).

C. Chemicals

1. Radiochemicals were purchased from The Radiochemical Centre, Amersham, Buckinghamshire, Great Britain.

2. Nucleosides and nucleotides were purchased from Schwarz Bioresearch Co., Orangeburg, N.Y., or from Boehringer Mannheim, GmbH, Mannheim, Germany.

3. Other chemicals were purchased from Sigma Chemical Co., St. Louis, Missouri, Serva Feinbiochemica GmbH & Co., Heidelberg, Germany, or from Difco Laboratories, Detroit, Michigan.

D. Abbreviations

EDTA ethylenediaminetetraacetic acid

EGTA ethyleneglycol-bis-(β-aminoethylether)-N,N'-tetraacetic acid

SDS sodium dodecyl sulfate

TCA trichloroacetic acid

NAD nicotinamide dinucleotide

dXTP deoxyribonucleoside triphosphate

III. METHODS

A. Standard Method [1]

1. Growth of Bacteria

Thymine-requiring bacterial strains are grown at the desired temperature in [^{14}C] thymine (2 μg/ml, 600 mCi/mole) containing Penassay broth [17.5 g Penassay broth (Difco) in 1000 ml H_2O] to a concentration of 2 x 10^8 cells/ml. The prelabel thus obtained is measured as acid precipitable radioactivity in a sample of known cell number. The prelabel is needed to correct for the slightly different cell numbers applied to the cellophane disks due to the inaccuracy in pipeting microliter amounts. The culture is cooled to 0°C. The bacteria are harvested, washed in buffer A (40 mM Tris·HCl pH 7.8, 10 mM EGTA, 20% w/v sucrose), and resuspended in the same medium at a density of 5 x 10^{10} cells/ml at 0°C. At this stage the bacteria should not be kept for more than 1 hr. Steps 2 and 3 are carried out at 0°-4°C.

2. Spheroplast Formation

One microliter of the cell suspension together with 1 μl lysozyme Brij solution (1 mg lysozyme/ml 1% Brij 58 in buffer A) are spread on a 1 cm^2 cellophane membrane disk previously placed on an agar plate A (2% Bacto agar in buffer A). Spheroplast formation occurs within 2 min. The preparation is unstable and should be used within 15 min. The spheroplasts are stable for a few hours if prepared on an agar plate A supplemented with 40% glycerol and kept at -20°C.

The activity of the system depends on the cell density. The optimal concentration range is 2 to 10 x 10^7 cells/cm^2 disk area.

3. Lysis and Concentration of the Lysate

The spheroplast-carrying cellophane disks are transferred to a cold agar plate B (2% Bacto agar in buffer B, 20 mM Tris HCl pH 7.8, 5 mM $MgCl_2$, 0.1 mM EDTA). The disks are either dried in a cold air stream lying on plate B or after transfer to a layer of filter paper soaked with 40% polyethylene glycol (PEG) 20,000 in buffer B, including 10 mM β-mercaptoethanol. Several disks can be transferred at a time if the lysate is prepared on small cellophane disks lying on a larger nitrocellulose filter (e.g., Sartorius filter SM 11 308, Sartorius, Göttingen, Germany). The filter carrying the disks is then transferred from plate A to plate B and subsequently to the filter paper soaked with PEG. After drying, the disks should display an opaque surface. No liquid must be left on the lysate after drying, since the activity of the system is very sensitive to dilution.

Lysis of the cells may be checked by the following criteria: (a) DNase (1 μg) added to a disk degrades parental DNA and inhibits DNA synthesis; (b) nucleosides and monophosphates are not efficiently incorporated; (c) the system is complementable (see Section III,C).

4. Incorporation of Precursors into Acid Insoluble Material

The standard reaction mixture in buffer B contains (μmoles/ml) KCl, 100; ATP, 1; deoxyribonucleoside triphosphates, 0.02, with one of them labeled with ^3H at 500 Ci/mole; one deoxyribonucleoside, corresponding to the labeled nucleotide, 0.2; NAD, 0.2. The addition of the deoxyribonucleoside prevents incorporation of labeled degradation products into unlysed cells. NAD is a necessary cofactor for polynucleotide ligase. One can block the ligase action by replacing NAD by 1 mM NMN. The triphosphates used have always been purchased from Schwarz or Boehringer (see Section II). Triphosphates from another source have given erratic results.

A drop (40-50 μl/1 cm^2 disk) of reaction mixture is pipeted onto Parafilm covering a metal plate kept at the desired temperature. The disk is transferred from agar plate B and floated on the drop. The system is covered with the lid of a petri dish to avoid evaporation. Care must be taken that no part of the lysate-carrying surface of the disk is flooded by the incubation mixture. Instead of the drop, a glass filter (Whatman GF/A) soaked with reaction mixture may be used. This technique is useful for all types of preincubation or chase experiments since flooding of the lysate is thus more easily prevented. For pulse experiments the reaction is started by shifting the disks from a filter soaked with incubation mixture at 0°C to a prewarmed drop, thus eliminating the time necessary for precursor equilibration.

5. Stopping the Reaction and Preparation of DNA

a. For Determination of Acid Insoluble Radioactive Material.

1. The reaction is stopped by putting the disk upside down (i.e., lysate side downward) onto a glass filter (Whatman GF/A) 2.4 cm in diameter, soaked with a mixture to stop the reaction (stopmix: 0.1 N NaOH, 0.2% SDS, 0.02 mg/ml calf thymus DNA, 2% saturated $Na_4P_2O_7$). The disk is removed and the filter is washed in 0.25 M TCA, 0.05 M TCA and methanol (10 min each), dried, and counted in a liquid scintillation counter.

2. Alternatively, the reaction can be stopped by putting the disk into 0.5 ml of fivefold concentrated stopmix and heating to 80°C. DNA is then precipitated by addition of 2 M TCA and collected on nitrocellulose filters. The heat treatment in the alkaline stop mixture destroys an acid precipitable compound generated when [^{3}H]dCTP is used as the labeled precursor.

b. For Further Studies on the Reaction Products (Density and Sedimentation Analysis). To stop the reaction, the disk is transferred to a solution containing 0.5% Sarkosyl and 100 mM EDTA, pH 8.0. Incubation with 1 mg/ml pronase for 12 hr at 60°C in this solution and removal of the DNA from the disk with the help of a glass rod or by shaking yields the native reaction product in solution. The denatured product is more easily prepared by placing the disk into a stop solution consisting of 0.2 N NaOH and 0.1 M EDTA.

B. Modification of the Standard Method

1. Mitomycin Treatment [5]

To prevent cellular DNA synthesis (which is necessary in studies of phage DNA replication and certain complementation experiments, see below), the cells are harvested and resuspended at a density of 2 x 10^9 cells/ml in PTM buffer (10 mM Tris·HCl, pH 7.4, 10 mM $MgCl_2$) at growth temperature and aerated for 10 min. Mitomycin is added and the suspension incubated for another 10 min in the dark without aeration. Finally, the cells are washed in cold PTM buffer. The amount of mitomycin necessary depends on the repair functions of the strain used. For strains derived from H 560, 50 μg/ml is used. The minimal necessary amount should always be used to avoid possible side effects.

2. EDTA-Lysozyme Lysis

Lysis of cells grown in a minimal medium is optimal only if the EGTA in plate A and buffer A is replaced by EDTA. Such preparations, unlike the standard ones, have been found to show a 50% reduction in DNA-synthesizing activity after RNase treatment [16].

C. Complementation Techniques

1. Mixed Lysates

For the preparation of mixed lysates, the concentrated cell suspensions in buffer A are mixed in the desired ratios. The lysates are then prepared as described.

2. Complementation with Macromolecules

Between 1 and 10 μl of a solution containing macromolecules may be added to a 1 cm^2 disk just after lysis. It is concentrated together with the lysate by one of the drying procedures. The solutions added should not contain a high concentration (> 0.4 M) of salt, since such an addition can irreversibly alter the system in spite of the fast diffusion of small molecules into the agar.

3. Preparation of Extracts

Cells are harvested, washed in 10 mM Tris HCl pH 8, 1 mM EDTA, 20% w/v sucrose, and resuspended in the same buffer in the cold. The minimal amount of buffer used is 1 ml per gram of wet cells. To a 1 ml cell suspension, 0.25 ml of lysozyme solution (lysozyme, 2 mg/ml in 120 mM Tris·HCl pH 8, 50 mM EDTA) is added and incubated for 3 min at 0°C. The cells lyse upon addition of 1.25 ml 1% Brij 58, 10 mM EDTA. The lysate is centrifuged at 60,000 *g* for at least 30 min in the cold. This method yields DNA-free supernatants. In the case of cells grown in rich medium, EDTA may be replaced by EGTA. The time of lysozyme action should then be increased to 10 min.

IV. GENERAL CHARACTERISTICS OF THE SYSTEM

A. Influence of Temperature on Incorporation Kinetics and the Stability of the System

The system, when prepared from wild-type cells, shows that DNA synthesis is linear with time for 30 min when incubated between 25° and 37°C. The reaction rate is independent of the incubation temperature in this temperature range. At higher temperatures the system breaks down, i.e., nonlinear kinetics are found and the rate decreases. With temperature-sensitive *dna* mutants, the permissive temperatures are generally lower in vitro than they are in vivo [2]. The individual strains vary and their in vitro characteristics have to be tested when they are used in the cellophane disk system.

B. Criteria for In Vitro DNA Replication

The DNA synthesis in the system reflects in vivo DNA replication by the following criteria.

1. DNA synthesis is semiconservative [1].

2. Short nascent intermediates ("Okazaki pieces") are formed and joined [1].

3. The synthesis is independent of DNA polymerase I (see Section IV,D).

4. The system shows sensitivity to replication-inhibiting drugs (nalidixic acid, mitomycin C) and UV irradiation [1].

5. The system is temperature sensitive when prepared from some temperature-sensitive *dna* mutants [2].

6. DNA synthesis is reduced in a system prepared from cells known to be inhibited in initiation of a new round of DNA replica-

tion, e.g., dna A mutants grown at nonpermissive temperature for one generation time [17] or cells treated with chloramphenicol for one generation time [1].

7. Methionine auxotrophic strains exhibit reduced DNA synthesis in vitro when tested after methionine starvation for one generation [18] just as they do in vivo [6].

C. Residual DNA Synthesis

A small amount of DNA synthesis is usually observed when the system is run under inhibitory conditions [1]. The nature of this type of synthesis has not yet been investigated.

D. DNA Synthesis Different from Replication

The addition of polymerase I to the cellophane system does not provoke additional DNA synthesis unless single strand breaks are introduced into the DNA. It has been observed that nonspecific stimulation occurs after the addition of extracts or commercial protein preparations such as bovine serum albumin, even in systems prepared from polymerase I-deficient strains. It is suspected that this stimulation is due to contaminant nuclease(s) in these preparations that yield(s) unspecific templates for other DNA polymerases present in the system.

V. APPLICATIONS

A. Bacterial Systems

1. Biochemical Characterization of dna-Mutants and of Replication Proteins [2,7]

Two classes of dna mutants have been distinguished phenotypically: (1) mutants deficient in initiation of a new round of replication (dna A and dna C) and (2) mutants inhibited in the synthesis of previously initiated DNA strands (dna B, dna E, and dna G) [3].

Mutants of the second class show temperature-sensitive DNA synthesis in the cellophane system. The synthesis at the nonpermissive temperature is stimulated in a mixed lysate prepared with mitomycin-treated wild-type cells or from two strains belonging to two different complementation groups. dna E and dna G mutants can also be complemented with extracts prepared from either wild-type cells or from mutants belonging to another complementation group. Since unspecific stimulation of dna mutants has been observed in the in vitro system, it is essential that controls always be run, showing that the complementation is specific. dna mutants of unknown genotype have been investigated by this method and found to belong to one of the three genetically defined groups mentioned above.

The complementation of temperature-sensitive lysates with fractionated wild-type extracts has served as an assay during the purification of the dna E and the dna G gene products [7].

The dna E gene product copurifies with DNA polymerase III activity [8-10]. In the case of the dna G gene product, it has been possible to purify the temperature-sensitive protein of the dna G mutant NY 73 (with 25° and 37°C as permissive and nonpermissive temperatures, respectively, in vitro) by testing it in a lysate of BT 308 (with 25°C as the nonpermissive temperature in vitro). Both wild-type and mutant protein have shown similar chromatographic behavior [17].

2. Identification of Replication Intermediates

DNA replication occurs discontinuously, i.e., the DNA single strands are synthesized in pieces that are subsequently joined by polynucleotide ligase to give a whole molecule. To study the intermediates and the joining reaction, the following techniques have been used:

a. Blocking of Polynucleotide Ligase. Nicotinamide mononucleotide (NMN) (1 mg/ml) is included in the incubation mixture (NMN can be chased by incubation of the system on a nonradioactive incorporation mixture containing 0.3 mg NAD/ml. After the reaction is stopped, the products can be analyzed by alkaline sucrose gradient centrifugation. It is essential to use bacteria with prelabeled chromosomes in order to have an internal control showing that the size of the products does not reflect nucleolytic degradation. The prelabeled DNA must be larger than the replication intermediates obtained in vitro before the chase.

Intermediates of two discrete size classes have been found [11]. The products of each class are thought to represent copies of one of the two opposite DNA strands [12].

b. Influence of Precursor Concentration [11]. Using the technique just described, it is possible to show that the size of the intermediates obtained varies with the precursor concentration. A fivefold reduction in dXTP concentration, resulting in a threefold reduction in the rate of synthesis, results in the accumulation of smaller intermediates in the presence of NMN.

B. Bacteriophage Systems

1. Conversion of ØX 174 DNA to Double-Stranded Molecules [13]

ØX 174 single-stranded DNA can be converted in the cellophane system to the replicative form containing a gap in the complementary strand (RF II) and, depending upon the presence of polynucleotide ligase and NAD, to the closed-circular replicative form (RF I). The cell-free system is prepared by spreading ^{32}P-labeled ØX DNA (10^8 molecules) together with 10^8 cells of E. coli H 560 onto 1-cm^2 cellophane disks and subsequent lysis and system preparation as described above. Addition of ØX DNA to the disks after lysis of the cells has given erratic results.

2. Replication of Bacteriophage λ DNA [4,7]

a. General Procedure. To study the replication of bacteriophage λ DNA in the cellophane disk system, prelabeled cells are infected at a multiplicity of input of four to eight per cell after mitomycin treatment. For the strain used in these studies, E. coli H 502, 20 μg/ml of mitomycin is used. Adsorption is allowed for 15 min at 37°C at a cell concentration of 5 x 10^9/ml in PTM buffer (see Section III,B,1) containing 3 mM KCN. After centrifugation, the cells are quickly resuspended in prewarmed, aerated Penassay Broth containing 20 μg/ml of thymine, and phage development is allowed in vivo for the desired time. The in vivo incubation is terminated by cooling in an ice-NaCl mixture, and the system is then prepared as described above with drying according to the PEG method. Air drying has given unreproducible results. The incubation temperature used in vitro is 25° to 28°C. Replication defective mutants (O^- or P^-) show little residual synthesis, which has been shown to be of bacterial origin. This residual synthesis is lower than that of systems prepared from uninfected cells treated identically before preparation of the system.

b. Dependence of the Product Obtained on the Length of Time of In Vivo Preincubation. Sedimentation analysis of the products obtained after different times of in vivo preincubation has been performed. In these experiments 4-cm^2 disks are used. It is found that the products obtained in vitro reflect the type of synthesis occurring in the early or late phase of λ replication. In vivo, at 37°C, incubation times of the infected cells of 5 to 7 min for early- and more than 10 min for late-phase replication (see Ref. 14 for review) are used in each case prior to the system preparation. A switch from early- to late-phase type replication in vitro has not been observed even upon prolonged in vitro synthesis.

c. In Vitro Complementation of λ O and P Gene Products. In vitro complementation of the O and P gene products has been shown in a mixed lysate of cells infected with either λ CI O125 or λCI P80 [4]. To distinguish the λ DNA synthesis obtained from residual bacterial synthesis (see Section 2,b), a hybridization technique has been used. The reaction is run on 15-cm^2 disks and stopped in 0.1 M EDTA, pH 7.8. The product is then treated with 20 μg/ml of RNase for 1 hr at 37°C, followed by Pronase treatment (see Section III,A). The DNA is extracted with phenol, extensively dialyzed, denatured, and hybridized to nitrocellulose filters carrying immobilized single-stranded λ DNA in the presence of excess single-stranded E. coli DNA [15].

d. Density Analysis of the Products. Density label experiments carried out according to the methods described in Section III,A with systems prepared from λ-infected cells, have yielded DNA fully substituted with bromouracil, indicating reinitiation of new rounds of replication of the λ chromosome in vitro. To prepare the DNA for the density gradient analysis, it has to be sheared by repeatedly squirting the solution through a syringe with a 0.55 mm Ø needle attached to it.

ACKNOWLEDGMENTS

We thank Peter Symmons for his help during the preparation of the manuscript and all of our colleagues who made unpublished data available to us.

The cellophane membrane system was conceived by Friedrich Bonhoeffer. Together with many of our colleagues he has developed numerous technical details and applications.

REFERENCES

1. H. Schaller, B. Otto, V. Nüsslein, J. Huf, R. Herrmann, and F. Bonhoeffer, J. Mol. Biol., 63, 183 (1972).

2. J. A. Wechsler, V. Nüsslein, B. Otto, A. Klein, F. Bonhoeffer, R. Herrmann, L. Gloger, and H. Schaller, J. Bacteriol., 113, 1381 (1973).

3. J. Gross, Current Topics Microbiol. Immunol., 57, 39 (1972).

4. A. Klein and A. Powling, Nature New Biol., 239, 71 (1972).

5. B. H. Lindqvist and R. L. Sinsheimer, J. Mol. Biol., 30, 69 (1967).

6. C. Lark, J. Mol. Biol., 31, 401 (1968).

7. A. Klein, V. Nüsslein, B. Otto, and A. Powling, DNA Synthesis in Vitro (R. D. Wells and R. B. Inman, eds.), University Park Press, Baltimore, Md., 1973, p. 185.

8. M. L. Gefter, Y. Hirota, T. Kornberg, J. A. Wechsler, and C. Barnoux, Proc. Natl. Acad. Sci. U.S., 68, 3150 (1971).

9. V. Nüsslein, B. Otto, F. Bonhoeffer, and H. Schaller, Nature New Biol., 234, 285 (1971).

10. B. Otto, F. Bonhoeffer, and H. Schaller, Europ. J. Biochem., 34, 440 (1973).

11. B. M. Olivera and F. Bonhoeffer, Nature New Biol., 240, 233 (1972).

12. R. Herrmann, J. Huf, and F. Bonhoeffer, Nature New Biol., 240, 237 (1972).

13. B. M. Olivera and F. Bonhoeffer, Proc. Natl. Acad. Sci. U.S., 69, 25 (1972).

14. D. Kaiser in The Bacteriophage Lambda (A. Hershey, ed.), Cold Spring Harbor Laboratory, Cold Spring Harbor, N.Y., 1971, p. 195.

15. E. Winocour, Virology, 34, 571 (1968).

16. K. G. Lark, personal communication.

17. V. Nusslein and A. Klein, unpublished data.

18. C. Lark, personal communication.

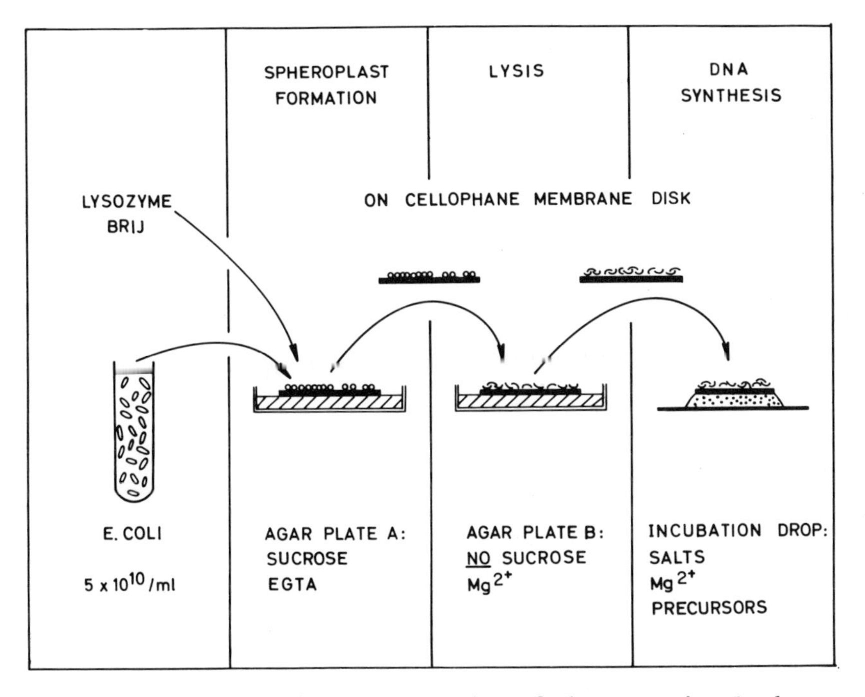

FIG. 1. Schematic representation of the steps involved in the preparation and use of the cellophane membrane system. The permeability of the membrane to small molecules and the thin layer of cells used allow a fast change of small molecular components (sucrose, EGTA, Mg^{2+}, precursors) during the transfers.

Chapter 9

DNA REPLICATION IN PLASMOLYZED *ESCHERICHIA COLI*

Reed B. Wickner*

Department of Developmental Biology and Cancer
Albert Einstein College of Medicine
Bronx, New York

I. INTRODUCTION

Cell permeabilization by treatment with concentrated sucrose was first demonstrated by Gros, Gallant, Weisberg, and Cashel [1] and has proven to be a useful method in the study of RNA and protein synthesis [2-4]. Cells treated in a similar manner also

*Present address: Laboratory of Biochemical Pharmacology, National Institutes of Health, Bethesda, Maryland.

carry out normal DNA replication [5] utilizing exogenous precursors. Other permeabilized cell preparations that carry out normal DNA replication have been described (Chapters 6 to 8 and 10, and Refs. 6-9).

II. PLASMOLYSIS OF E. COLI

A. Bacterial Strains

E. coli D110 is a *pol* A1, *endo* A^-, thy^- strain [7] and has been used routinely. End^+ strains also showed almost exclusively replicative synthesis, but *pol* A^+ strains carried out a large amount of nonreplicative synthesis. Other strains that have been used include *E. coli* BT1029 (*pol* A_1, end^-, thy^-, *dna* Bts) [10], *E. coli* BT1026 (*dna* Ets, *pol* A_1, end^-, thy^-) [10], *E. coli* NY73 (*pol* A_1, thy^- *dna* G3, rif^r, str^r, leu^-, *met* E) [11,12], and *E. coli* PC22 (*pol* A_1, *his*, str^r, *arg*, *mtl*, *dna* C2ts) [11,12].

B. Media

DC medium contains, per liter, 10 g of casamino acids, 10 g of glucose, 0.1 g of $MgSO_4$, 5 g of NaCl, 1 g of NH_4Cl, 10 mg of yeast extract, 3.5 g of K_2HPO_4, and 1.5 g of KH_2PO_4.

TG medium [8] contains 0.5 g of NaCl, 8 g of KCl, 1.1 g of NH_4Cl, 0.2 g of $MgCl_2$, 12.2 g of Tris base, and 0.8 g of pyruvic acid per liter. The pH is adjusted to 7.4 before autoclaving. After autoclaving, to each liter is added 1 ml of 0.16 M Na_2SO_4, 2 ml of 1 M $CaCl_2$, 1 ml of 10^{-4} M $FeCl_2$, 5 ml of 40% glucose, and 5 ml of thiamine (2 mg/ml).

C. Plasmolysis Procedure

All operations were carried out as rapidly as possible. Cells were routinely grown in DC medium supplemented with 2 μg/ml of [^{3}H]thymine (2 Ci/mole) with vigorous aeration. Cells were harvested in log phase (usually at about OD_{650} = 0.5) by chilling the culture on ice (<1 min) and centrifuging for 2 min at 10,000 rpm in the GSA rotor of a Sorvall centrifuge at 2°C. An aliquot of the culture is used to measure the OD_{650} and thereby the cell concentration. A second aliquot is digested in alkali and the DNA is precipitated with trichloroacetic acid (TCA) and counted as described below. Cells were quickly suspended in ice-cold TG medium containing 2 μg/ml of thymine and centrifuged at room temperature at 13,000 g for 2 min. The pelleted cells were finally suspended by vortexing in cold 0.04 M Tris chloride, pH 8.0, 0.01 M ethyleneglycol-bis-(β-aminoethyl ether) N,N'-tetraacetic acid, and 2 M sucrose at a concentration of about 10^{11} cells/ml. Cells may be stored at this stage at -20°C for at least 3 months without appreciable loss in their in vitro DNA replicative activity.

For the measurement of in vitro DNA synthesis, reaction mixtures (0.05 ml) contain 10 mM dithiothreitol, 100 mM KCl, 50 mM morpholinopropane sulfonic acid, pH 7.5, 2 mM ATP, 10 mM $MgCl_2$, 0.1 mM EDTA, 40 μM each of dATP, dGTP, and dCTP, and 20 μM [α-^{32}P]dTTP (200-500 cpm/pmole). Between 5 x 10^6 and 10^9 (usually about 5 x 10^8) cells are placed in assay tubes on ice; 50 μl of the reaction mixture is added, the cells are gently suspended, and the tubes are placed in a 37°C water bath for 15 min. The reactions are stopped by adding 0.5 ml of 0.5 N NaOH - 1% sodium dodecyl sulfate and 0.2 ml of 0.1 M sodium pyrophosphate containing 1 mg/ml of calf thymus DNA. After 15 min at 80°C, the tubes are filled with 4 ml of 10% trichloroacetic acid, and acid-insoluble material is collected on glass-fiber filters and counted.

Labeling the cellular DNA in vivo (with ^3H) allows one to correct in vitro incorporation values (in ^{32}P) for variations in the number of cells per assay due to pipeting errors, etc. It also allows one to check cell permeability by adding DNase I (10 μg) to a standard incubation. More than 95% of the parental DNA should be degraded to trichloroacetic acid-soluble material during a 15-min incubation. The proportion degraded may represent the proportion of the cells made permeable by sucrose treatment.

It is also necessary to determine, in a given experiment, whether the DNA synthesis one is measuring is replicative or repair in nature. This has been emphasized in the case of toluenized cells (Chapter 7 and Ref. 7). This is most conveniently done by checking whether nalidixic acid (100 μg/ml) inhibits the synthesis. Of course, no single criterion of replicative synthesis is definitive.

III. PROPERTIES AND USES OF PLASMOLYZED CELLS

Cells treated with high concentrations of sucrose become permeable to small molecules [1-5] as well as to some macromolecules [5]. Cell viability drops to about 1%, and protein, RNA, and DNA synthesis takes place intracellularly utilizing exogenous precursor amino acids and triphosphates [1-5]. The RNA and protein synthesis exhibited by plasmolyzed cells remains under many of the normal physiologic controls [1-4].

DNA synthesis in plasmolyzed cells [5] is semiconservative; requires ATP, as is true in other DNA replication systems [6-9]; and is temperature sensitive in plasmolyzed cells of *dna* B, *dna* E, or *dna* G ts mutants (immediate-stop type) but not in *dna* A or *dna* C mutants (initiation defective). If *dna* A ts or *dna* C ts cells are grown at 30° (the permissive temperature) and then incubated at 42° (nonpermissive temperature) for 60 min to allow completion of rounds of replication, plasmolysis yields preparations nearly inactive in DNA synthesis. Similar results are obtained following prolonged chloramphenicol treatment [5]. These results are those expected of DNA replication in a system defective in the in vitro initiation of rounds of replication. (For a review, see Ref. 13.)

The DNA synthesis is also sensitive to nalidixic acid, a specific inhibitor of replicative DNA synthesis [14]. DNA pulse labeled in vitro resembles Okazaki pieces [15] and can be chased into larger material.

Recently, Wovcha, Tomich, Chiu, and Greenberg have studied DNA replication in T4-infected *E. coli* made permeable by a modification of the method described here [16]. The DNA synthesis was shown to depend on the T4 genes 1, 32, 41, 42, 43, 44, and 45 when exogenous deoxyribonucleotides were the DNA precursors.

Staudendauer and Hofschneider [17] have shown that M_{13}-infected plasmolyzed cells can initiate rounds of RF replication.

REFERENCES

1. F. Gros, J. Gallant, R. Weisberg, and M. Cashel, *J. Mol. Biol.*, *25*, 555 (1967).

2. J. Gallant and M. Cashel, *J. Mol. Biol.*, *25*, 545 (1967).

3. M. Cashel and J. Gallant, *J. Mol. Biol.*, *34*, 317 (1968).

4. F. Ben-Hamida and F. Gros, *Biochimie*, *53*, 71 (1971).

5. R. B. Wickner and J. Hurwitz, *Biochem. Biophys. Res. Commun.*, *47*, 202 (1972).

6. H. P. Vosberg and H. Hoffmann-Berling, *J. Mol. Biol.*, *58*, 739 (1971).

7. R. Moses and C. C. Richardson, Proc. Natl. Acad. Sci. U.S., 67, 674 (1970).

8. H. Schaller, B. Otto, V. Nusslein, J. Huf, R. Herrmann, and F. Bonhoeffer, J. Mol. Biol., 63, 183 (1972).

9. W. T. Wickner, D. Brutlag, R. Schekman, and A. Kornberg, Proc. Natl. Acad. Sci. U.S., 69, 965 (1972).

10. J. A. Wechsler, V. Nusslein, B. Otto, A. Klein, F. Bonhoeffer, R. Herrmann, L. Gloger, and H. Schaller, J. Bacteriol., 113, 1381 (1973).

11. P. L. Carl, Mol. Gen. Genetics, 109, 107 (1970).

12. M. L. Gefter, Y. Hirota, T. Kornberg, J. Wechsler, and C. Barnoux, Proc. Natl. Acad. Sci. U.S., 68, 3150 (1971).

13. D. W. Smith, Progress in Biophysics and Molecular Biology, Vol. 26 (J. A. V. Butler and D. Noble, eds.), Pergamon Press, New York, pp. 321-408.

14. W. A. Goss, W. H. Deitz, and T. M. Cook, J. Bacteriol., 88, 1112 (1964).

15. R. Okazaki, T. Okazaki, K. Sakabe, K. Sugimoto, R. Kainuma, A. Sugino, and N. Iwatsuki, Cold Spring Harbor Symp. Quant. Biol., 33, 129 (1968).

16. M. G. Wovcha, P. K. Tomich, C.-S. Chiu and G. R. Greenberg, Proc. Natl. Acad. Sci. U.S., 70, 2196 (1973).

17. W. L. Staudendauer and P. H. Hofschneider, Biochem. Biophys. Res. Commun., 54, 578 (1973).

Chapter 10

A SOLUBLE-ENZYME EXTRACT FOR THE STUDY OF DNA REPLICATION

Joel H. Weiner, Douglas Brutlag, Klaus Geider,*
Randy Schekman, William Wickner, and Arthur Kornberg

Department of Biochemistry
Stanford University Medical Center
Stanford, California

*Present address: Max-Planck Institut für Medizinische Forschung, Abteilung Molekulare Biologie, Heidelberg, West Germany.

I. INTRODUCTION

Gentle lysis of E. coli, followed by high-speed sedimentation yields a supernatant relatively low in nucleic acids and containing enzymes that support DNA replication. This extract converts phage ØX174 and M13 single-stranded DNAs to the duplex replicative forms and thus provides a useful system for analyzing this stage of their DNA replication. The lysate serves as a source for identifying the components of the systems responsible for replication of the more complex phages. This method differs from the usual lysozyme lysis procedures in two ways: (a) omission of EDTA, and (b) inclusion of a brief temperature shift, usually from 4° to 30°C.

II. MATERIALS AND METHODS

A. Strains

E. coli H560 is F^+, pol A_1 (to reduce the background of DNA synthesis), end(onuclease)1^- (to reduce DNA degradation), and thy^-. The strain was provided by Dr. H. Hoffmann-Berling. Also used were: E. coli SK108, a strain that lacks polynucleotide phosphorylase (to reduce the background of RNA synthesis), pol A_1, and RNase I^-; strains carrying conditional lethal mutations in genes required for *E. coli* chromosome replication, e.g., *E. coli* *dna* A⁻, *dna* B⁻, etc.; and lambda lysogens that had been induced.

The procedure to be described has been optimized for studying the conversion of ØX174 and M13 single strands to the replicative form. The procedure may have to be varied, especially with strains that are more resistant to lysozyme lysis.

B. Growth of Cells

Cells were inoculated at an initial low density in Hershey broth (8 g nutrient broth, 5 g bactopeptone, 1 g glucose, 5 g sodium chloride per liter) and grown in a 100-liter fermentor at 37°C with aeration but no agitation (to avoid foaming) to early log phase (OD_{600} = 0.5). Antifoam was not used because it inhibited the ØX replication activity of the lysate and interfered with ammonium sulfate fractionation. Cells were collected by centrifugation in a refrigerated Sharples centrifuge, but the culture was not cooled before collection. The cell paste (200 g/liter) was suspended in 0.05 M Tris-HCl, pH 7.5 - 10% sucrose to a level of 5-10 x 10^{10} cells/ml (a 400-fold concentration of the harvested culture) and quick frozen in liquid nitrogen in polyallomer tubes or Erlenmeyer flasks that were stored at -20°C. Such cell pastes have been stable for a year or more.

Unlabeled ribonucleotides and deoxyribonucleotides were obtained from P-L Biochemicals; [α^{32}P]deoxynucleoside triphosphates were synthesized by the procedure of Symons [1]. Lysozyme was obtained from Worthington Biochemical Corp. Single-stranded M13 and ØX174 DNAs were extracted by the method of Ray [2] or the phenol-cresol extraction procedure of Kirby et al. [3].

C. LYSIS

Frozen cells were thawed slowly in a water bath at 4°C. The cells cannot be efficiently lysed without the freezing step before lysozyme treatment. To the suspension were added with stirring 0.025 vol of lysozyme (4 mg/ml in 0.05 M Tris·HCl, pH 7.5 - 10% sucrose) and 4 M NaCl to a final concentration of 0.1 M. The suspension was kept on ice for 30 min; but this time was extended for strains that did not lyse well. The suspension was then warmed for 60-90 sec in a water bath set at 37°C. This is a critical step. Warming allowed the suspension to reach 30°C and was carried out in 30-ml tubes to ensure more uniform warming. The lysate was then immediately cooled in an ice bath. To avoid this exposure of extracts with thermosensitive proteins, shifts were made in a bath at 25° to 30°C or a nonionic detergent, 0.5% Brij 58, was used. However, possible interference by the detergent with the purification of enzymes must be checked.

The lysate was centrifuged in polycarbonate tubes at 50,000 to 250,000 *g* at 0°C for 30 to 90 min. The duration and speed were determined by the volume of extract. The supernatant fraction is a clear, amber nonviscous solution. It has been stable for one day after storage at 0°C; quick frozen in liquid nitrogen and stored at -90° or -20°C, it has been stable for 1 year. Repeated freezing and thawing results in loss of activity. Nucleic acids in the supernatant may be removed by addition of DE52 DEAE-cellulose (equilibrated in 0.05 M Tris·HCl, pH 7.5 - 10% sucrose - 0.15 M NaCl; a 15-ml packed volume was used for 100 ml of extract). The final suspension, separated from the DEAE-cellulose by filtration, contains about 20 mg of protein/ml. Phase partition in polyethylene glycol-dextran with 3 M NaCl may also be used to remove nucleic acids [4]. Slight modifications of the above procedure are described in Chapters 11a and 11b.

III. APPLICATIONS OF THE SOLUBLE-ENZYME EXTRACT

A. Conversion of M13 Single Strands to Their Duplex Form--A Rifampicin-Sensitive Replication System

M13, a small filamentous bacteriophage, exploits a replication system of *E. coli* distinct from that responsible for host chromosome replication. DNA synthesis on an M13 template is inhibited by rifampicin, an inhibitor of RNA polymerase, a finding that indicates that RNA synthesis plays a role in DNA synthesis. The details for the in vitro assay of DNA synthesis on an M13 template have been published [5]. The product of the reaction is a duplex replicative form with a nearly full length linear complementary strand; the gap in the duplex is located at a unique place relative to the viral template [6].

B. Conversion of ØX174 Single Strands to Their Duplex Form--A Rifampicin-Resistant Replication System

The conversion of ØX174 SS DNA to the duplex form also proceeds via RNA initiation but does not depend on RNA polymerase and is not sensitive to inhibition by rifampicin. ØX replication requires a number of *E. coli* replication proteins: *dna* A, *dna* B, *dna* C-D, *dna* E, and *dna* G [7,8], suggesting that ØX may be exploiting the *E. coli* chromosome replication system and probably a novel RNA synthetic step. The optimal concentration of assay components for the in vitro conversion have been published [8]. The product of this reaction is also a full-length linear complementary strand on a circular template strand.

C. Resolution and Reconstitution of the Replication System

1. The DNA Polymerase III Star System [9]

Two proteins necessary for both ØX174 and M13 duplex DNA synthesis have been purified from the soluble extract. One, DNA polymerase III star (pol III*) appears to be a complex and more labile form of DNA polymerase III. The second protein, copolymerase III* (copol III*), is required for the activity of pol III* on long stretches of single-stranded DNA. Both pol III and pol III* contain the product of the *dna* E gene. On DNA templates with short gaps, both pol III and pol III* are stimulated by 10% ethanol and inhibited by KCl and a sulfhydryl blocking agent; copol III* acts only with pol III*.

2. Identification and Purification of Individual Components of the M13 and ØX174 Systems

When soluble extracts are prepared from *E. coli* strains carrying temperature-sensitive mutations of DNA replication (*dna* A, *dna* B, *dna* C, *dna* D, and *dna* G) ØX174 duplex synthesis but not M13 synthesis is temperature sensitive [7,8]. One exception is *dna* E, which is required for both. It is possible to complement a temperature-sensitive extract at restrictive temperature with fractions prepared from other mutants or wild-type cells and reconstitute ØX duplex synthesis. The *dna* G protein has been highly purified (Ref. 10 and Chapters 11a and 11b) and the other *dna* gene products have been partially purified with this complementation assay.

D. Resolution and Reconstitution of Other Replication Systems

1. Phage M13. Multiplication of the duplex replicative form [11] and its conversion to viral single strands [12] have been demonstrated and partially resolved with the soluble enzyme extract.

2. Phage lambda. The soluble lysate has been used to study in vitro assembly of bacteriophage lambda heads [13]. In this study it was necessary to alter lysis conditions slightly to prepare soluble extracts of lambda lysogens that had been induced.

3. Phage T7. Gentle lysozyme lysis of T7-infected cells has been used to prepare an extract capable of replication of this phage. It has been possible to purify the T7 gene-4 protein by a complementation assay [14].

4. Mutant polymerases. The soluble extract has been used to purify mutant DNA polymerase I proteins [15] and should be applicable to systems where a low level of DNA in the extract is desirable.

REFERENCES

1. R. H. Symons, Biochem. Biophys. Acta, 190, 548 (1969).

2. D. S. Ray, J. Mol. Biol., 43, 631 (1969).

3. K. S. Kirby, E. Fox-Carter, and M. Ghest, Biochem. J., 104, 258 (1967).

4. T. Okazaki and A. Kornberg, J. Biochem., 239, 259 (1969).

5. W. Wickner, D. Brutlag, R. Schekman, and A. Kornberg, Proc. Natl. Acad. Sci. U.S., 69, 965 (1972).

6. H. Tabak, J. Griffith, K. Geider, H. Schaller, and A. Kornberg, J. Biol. Chem., 249, 3049 (1974).

7. R. B. Wickner, M. Wright, S. Wickner, and J. Hurwitz, Proc. Natl. Acad. Sci. U.S., 69, 3233 (1972).

8. R. Schekman, W. Wickner, O. Westergaard, D. Brutlag, K. Geider, L. L. Bertsch, and A. Kornberg, Proc. Natl. Acad. Sci. U.S., 69, 2691 (1972).

9. W. Wickner, R. Schekman, K. Geider, and A. Kornberg, Proc. Natl. Acad. Sci. U.S., 70, 1764 (1973).

10. S. Wickner, M. Wright, and J. Hurwitz, Proc. Natl. Acad. Sci. U.S., 70, 1613 (1973).

11. O. Westergaard, D. Brutlag, and A. Kornberg, J. Biol. Chem., 248, 1361 (1973).

12. J. Griffith, O. Westergaard, and A. Kornberg, unpublished results.

13. A. D. Kaiser and T. Masuda, Proc. Natl. Acad. Sci. U.S., 70, 260 (1973).

14. D. Hinkle and C. C. Richardson, personal communication.

15. D. Uyemura and I. R. Lehman, unpublished results.

Chapter 11A

ISOLATION OF dna GENE PRODUCTS OF ESCHERICHIA COLI

Sue Wickner, Michel Wright, Ira Berkower, and Jerard Hurwitz

Department of Developmental Biology and Cancer
Division of Biological Sciences
Albert Einstein College of Medicine
Bronx, New York

I. INTRODUCTION

Mutants of *E. coli* have been isolated that are temperature sensitive (ts) for DNA replication; the genes involved have been designated *dna* A, B, C, D, E, F, and G [1-3]. The product of the *dna* E gene is DNA polymerase III [4,5], while that of the *dna* F gene is ribonucleotide reductase [6]. *In vitro* DNA synthesizing systems have been developed that depend on one or more of the products of these *dna* genes [7-12] (see Chapters 6-10). One of these systems, in which crude extracts of *E. coli* catalyze the conversion of ØX174 single-stranded circular DNA to the double-stranded replicative form (see Chapter 10) depends on the products of *dna* B, C, D, E, and G genes [11-13]. This has been shown by the increased thermolability of ØX174 DNA-dependent dNMP incorporation in extracts from ts cells as compared with extracts from temperature-resistant revertant cells. The stimulation of inactive crude extracts of dna ts cells by fractions from wild-type or the other ts cells has provided complementation assays for the purification of the dna gene products. These complementation assays have been used to isolate the dna B [13], dna C (D) [14], dna E [15], and dna G [15] gene products. The dna B, C (D), E, and G gene products have been isolated from dna B, C, E, and G ts cells, respectively. In each case the gene product from ts cells was more thermolabile than the wild-type product, thereby establishing the identity of the protein purified. Assaying for dna E gene product by measuring DNA-dependent dNMP incorporation, Kornberg and Gefter [16] and Otto et al. [17] have purified this gene product. The dna E gene product has also been purified using a complementation assay employing lysates of E. coli concentrated on cellophane disks [5,17]. This assay has also been used to purify an activity that stimulates DNA synthesis by lysates of dna G ts cells at the nonpermissive temperature [18].

II. METHODS USED FOR PURIFICATION OF dna GENE PRODUCTS

A. Bacterial Strains

The following E. coli strains were used: HMS-83 (pol A_1, pol B_1, thy, lys) [19]; PC22 (pol A_1, his, str^r, arg, mtl, dna C2 ts) [3]; PC79 (pol A_1, his, str^r, mtl, dna D7 ts) [3]; NY73 (pol A_1, thy, leu, metE, rif^r, str^r, dna G3 ts) [4]; BT1029; (pol A_1, thy, endo I, dna B ts) [20], and BT1040 (pol A_1, endo I, thy, dna E ts) [4,20].

B. DNA

ØX174 DNA was prepared by the method of Sinsheimer [21] or Franke and Ray [22].

C. Growth of Cells

E. coli strain HMS-83 used for the purification of *dna* B, E, and G gene products was obtained commercially from Truett Laboratories. *E. coli* strain HMS-83 used for the purification of *dna* C (D) and ts strains used for the *dna* gene product complementation assays were prepared as follows. Cells were grown in a 100-liter fermenter without antifoam to an OD_{595} of 0.5 to 1.0 at 25 or 30°C in Hershey broth (5 g of NaCl, 5 g of Bacto-Peptone, 10 g of nutrient broth, and 1 g of glucose per liter) supplemented with 10 μg/ml of thiamine and 20 μg/ml of thymine. Cells were collected as quickly as possible (within 1 hr) by centrifugation at room temperature and resuspended at room temperature in a volume of 10% sucrose containing 0.05 M Tris·HCl, pH 7.5 equal to the weight of the cells. The cell suspension was rapidly frozen in a dry ice-ethanol bath and stored at -20°C until use (see also Chapter 10).

D. Preparation of ts Receptor Fractions for Use in Complementation Assays

1. Receptor Crude Extract

Cells grown and frozen in sucrose as described above were thawed in an ice-water bath, incubated with 0.2 mg/ml lysozyme and 0.2% Brij 58 for 30 min at 0°C, and centrifuged for 30 min at 100,000 *g* at 4°C. After centrifugation, the supernatant (10 to 20 mg protein/ml) was frozen in small portions and is designated receptor crude extract (see also Chapter 10).

2. Receptor Ammonium Sulfate Fraction

Cells grown, frozen, and thawed as above were lysed with 0.2 mg/ml lysozyme, 0.01 M Tris·HCl, pH 8.5, 10^{-3} M dithiothreitol,

0.02 M EDTA, 0.1% Brij 58, and 0.15 M KCl for 20 min at 0°C and centrifuged for 40 min at 50,000 g at 4°C. The supernatant was adjusted to a final concentration of 4% with a solution of 20% streptomycin sulfate and centrifuged at 10,000 g for 5 min. The supernatant was adjusted to 40% saturation with saturated, neutralized ammonium sulfate (at 4°C) and centrifuged at 10,000 g for 10 min. The pellet was resuspended in 0.1 vol (of original crude extract) of 0.01 M Tris·HCl, pH 7.5, 10^{-3} M dithiothreitol, 5 x 10^{-4} M EDTA, and 15% glycerol and dialyzed against the same buffer for 2 hr, thus reducing the salt concentration to 0.1 M. The receptor ammonium sulfate fraction (50 mg protein/ml) was then frozen in small portions. Both the receptor crude extracts and the receptor ammonium sulfate fractions have been stored frozen for at least 6 months without detectable change in their properties; repeated freezing and thawing, however, destroys activity.

E. Complementation Assays

Each assay (0.05 ml) contained 20 mM Tris·HCl, pH 7.5, 10 mM $MgCl_2$, 4 mM dithiothreitol, 0.04 mM each of dATP, dGTP, dCTP, and [α-^{32}P]dTTP (200-500 cpm/pmole), 10 μg/ml rifampicin, 5 mM ATP, 2.5 mM spermidine·HCl, 500 pmoles ØX174 DNA, 0.01 ml of receptor crude extract prepared from *E. coli* strains BT1029, PC22, PC79, or NY73, 5 μl of receptor ammonium sulfate fraction prepared from the same strain as the receptor crude extract, and protein fractions as indicated. The receptor ammonium sulfate fractions prepared from strains BT1029 and NY73 were inactivated by heating 10 min at 38°C immediately prior to use. The receptor fractions prepared from PC79 and PC22 were inactive as isolated. Assay mixtures were incubated at 25°C for 20 min and acid-insoluble radioactivity was measured.

One unit of dna gene product activity incorporated 1 nmole of dTMP under the conditions described. Specific activity refers to units of activity per milligram of protein; the latter was measured by the method of Bucher [23]. The complementing activity of the purified dna gene products varied with the preparation of receptor crude extract and receptor ammonium sulfate fraction used. In all cases, it is essential that the concentration of protein be as high as possible. Activity is very sensitive to dilution.

dTMP incorporation required ØX174 DNA, ATP, Mg^{2+}, purified dna B, C (D), or G gene product, receptor ammonium sulfate fraction, and receptor crude extract prepared from dna B, C (D), or G ts cells, respectively. The requirement for both the receptor ammonium sulfate fraction and the receptor crude extract was not absolute; large amounts of ammonium sulfate fraction would substitute for the crude extract requirement, and similarly, large amounts of crude extract would substitute for the ammonium sulfate fraction. However, the combination, as described, gave the most reproducible results. The addition of spermidine stimulated the reaction but was not always an absolute requirement. Rifampicin had no effect on the reaction but was added to all reaction mixtures to avoid any role of DNA-dependent RNA polymerase. The system was not dependent on dNTPs, presumably due to their presence in the receptor crude extracts. The assay was salt, glycerol, and sucrose sensitive; 0.05 M KCl inhibited the reaction 50%; 10% glycerol (final) inhibited 50%; 7.5% sucrose (final) inhibited 50%.

Under the conditions of the complementation assays, the rate of dTMP incorporation was linear between 10 and 40 min of incubation and was directly proportional to the amount of dna gene product added up to 0.035 U. The DNA product formed in each of the complementation assays sedimented in neutral sucrose gradients as an RF II structure and in alkali as full-length linear strands.

The complementation assay used here for dna gene product detection measures the activation of inactive crude extracts prepared from dna ts mutants. It is assumed that the most limiting and most easily inactivated component in the receptor crude extract is the dna ts protein. This assumption can only be rigorously proven by purification of the dna gene product from ts cells, and the demonstration of its temperature sensitivity compared with that purified from wild-type cells. The dna B, C (D), E, and G gene products have been isolated from dna B, C, E, and G ts cells, respectively [13-15], and in each case the ts gene product was more thermolabile than the wild-type product, thereby establishing the identity of the protein purified.

III. SEPARATION OF dna GENE PRODUCTS

As previously reported from this laboratory [12,15], the dna gene products of E. coli that can be measured by the ØX174 DNA-dependent complementation assays have been separated from each other. As shown in Fig. 1, resolution of most of the dna gene products can be obtained by DEAE-cellulose chromatography. Ammonium sulfate I fractions prepared as described below for the purification of dna C gene product and chromatographed on DEAE-cellulose columns reproducibly yielded the elution pattern as indicated. dna C activity almost completely passed through the column; next dna G and dna E (measured as DNA polymerase III) activities appeared; lastly, dna B activity eluted. In each case, 50% or more of each activity applied to the column was recovered. dna G and dna E are resolved by DEAE-Sephadex chromatography [15] or by sedimentation through glycerol gradients. Thus, each can readily be freed of contamination by the others [13]. The insert in Fig. 1 represents rifampicin-resistant, ØX174 DNA-dependent dNMP incorporating activity catalyzed by eluted fractions in the absence of receptor crude extracts. Only a small percentage (1%) of this activity was recovered from the starting ammonium sulfate fraction. If a stable complex of the *dna* gene products exists, only a small amount of this activity (indicated in the insert of Fig. 1) survives the gradient elution.

IV. PURIFICATION OF *dna* B GENE PRODUCT

The *dna* B gene product has been purified by modifying the procedure of Wright et al. (Ref. 13 and Chapter 11B, this volume).

A. Crude Extract and Streptomycin Sulfate Fraction

All purification steps were carried out at 4°C. *E. coli* HMS-83 (1 kg), suspended in 1 liter of 0.02 M potassium phosphate, pH 7.5, 0.1 M KCl, 1 mM EDTA, and 1 mM dithiothreitol, was disrupted by passage through a Manton-Gaulin laboratory homogenizer at 10,000 psi. The crude extract (2 liters) was adjusted to a final concentration of 4% with freshly prepared 20% streptomycin sulfate, mixed, and centrifuged at 100,000 *g* for 60 min.

B. Ammonium Sulfate Precipitation

The streptomycin sulfate supernatant (1500 ml) was adjusted to 40% saturation with solid ammonium sulfate (22.6 g/100 ml). The precipitate was collected by centrifugation at 10,000 *g* for 15 min, dissolved in 100 ml of 0.01 M potassium phosphate, pH 7.5, 1 mM EDTA, 1 mM dithiothreitol, and 20% glycerol (Buffer A), and dialyzed for 5 hr against the same buffer.

C. DEAE-Cellulose I Column Chromatography

The dialyzed fraction was diluted with buffer A to a final salt concentration of 0.1 M and was applied (1300 ml) to a 700 ml DEAE-cellulose column equilibrated with buffer A containing 0.1 M KCl. Salt concentrations were measured by conductivity relative to a KCl standard. The column was washed with 2 liters of buffer A containing 0.2 M KCl. dna B activity was eluted with buffer A containing 0.3 M KCl; 16-ml fractions were collected. Active fractions were pooled (1200 ml) and adjusted to 50% saturation with solid ammonium sulfate. The precipitate was collected by centrifugation, dissolved in 10 ml of buffer A, and dialyzed against buffer A for 4 hr.

D. DEAE-Cellulose II Column Chromatography

The dialyzed sample was diluted to 40 ml with 0.02 M potassium phosphate, pH 6.5, 1 mM dithiothreitol, 1 mM EDTA, and 20% glycerol (buffer B) and applied to a 25 x 3 cm column of DEAE-cellulose equilibrated with buffer B. The column was developed with a 1-liter linear gradient from 0.1 to 0.5 M KCl in buffer B; 12-ml fractions were collected. The dna B activity eluted at 0.3 M KCl; active fractions were pooled (250 ml) and dialyzed for 3 hr against 0.02 M Tris·HCl, pH 8.5, 0.05 M KCl, 1 mM EDTA, 1 mM dithiothreitol, and 20% glycerol (buffer C).

E. DEAE-Sephadex Column Chromatography

The dialyzed fraction was applied to a 2 x 20 cm column of DEAE-Sephadex equilibrated with buffer C. The column was developed with a 500-ml linear gradient from 0.2 to 0.7 M KCl in buffer C; 3.5-ml fractions were collected. dna B activity was eluted with

0.35 M KCl. Active fractions were pooled (100 ml), dialyzed against buffer A for 2 hr, and applied to a 15-ml DEAE-cellulose column equilibrated with buffer A. dna B activity was eluted with buffer A containing 1 M NaCl. This fraction (8 ml) was stored frozen.

F. Glycerol Gradient Centrifugation

A portion of the DEAE-Sephadex fraction was dialyzed for 30 min against 0.1 M KCl, 0.02 M potassium phosphate buffer, pH 7.5, 1 mM EDTA, 1 mM dithiothreitol, 1 mM $MgCl_2$, and 5% glycerol. The dialyzed sample (0.2 ml) was layered on a 5-ml linear gradient of 15 to 35% glycerol in the same buffer and centrifuged in the Spinco SW 50.1 rotor at 50,000 rpm for 7 hr. The dna B activity sedimented through one-third of the gradient. This fraction was unstable, possibly due to its low protein concentration. The final preparation retained 35% of its activity after 5 days.

The purification and yield of dna B activity are summarized in Table 1.

G. Properties of dna B Gene Product

The glycerol gradient fraction was analyzed by polyacrylamide gel electrophoresis using the system of Brown [24]. As shown in Fig. 2, there was one major protein band, and this corresponded to the dna B complementing activity. The activity purified from dna B ts cells was thermolabile when compared with that purified from wild-type cells, thus proving that the dna B gene product had been isolated. The dna B gene product has a molecular weight of about 250,000 as determined by glycerol gradient sedimentation, and its complementing activity is unaffected by treatment with

TABLE 1

Purification of dna B Gene Product

Fraction	Total U	Total protein, mg	Specific activity, U/mg	% Recovery
Crude extract[a]	--	73,000	--	--
Ammonium sulfate	13,000	6,900	2	100
DEAE-cellulose I	5,800	124	45	45
DEAE-cellulose II	3,500	38	930	27
DEAE-Sephadex	1,200	0.75	1,600	10
Glycerol gradient[b]	1,200	0.07	16,400	10

[a]Crude extracts prepared with a Manton-Gaulin homogenizer could not be assayed directly for dna gene products because dTMP incorporation was independent of added ØX174 DNA and receptor crude extracts. However, when cells were lysed by the Brij-lysozyme method described for the preparation of receptor crude extracts, the activity was dependent on ØX174 DNA and receptor crude extract; close to 100% of each dna gene product-complementing activity in such crude extracts could be recovered in the 0% to 40% ammonium sulfate fraction with a 10- to 20-fold purification. The recovery of activity in the ammonium sulfate fraction per gram of cells was nearly the same whether cells were lysed with Brij-lysozyme or by passage through a Manton-Gaulin homogenizer.

[b]This step was performed with only part of the DEAE-Sephadex fraction. The values reported assume that the yield and purification would be the same if the entire fraction were subjected to the glycerol gradient procedure.

N-ethylmaleimide. Fractions purified through the DEAE-cellulose II step were free of dna C (D) and G gene products (measured by ØX174 DNA-dependent complementation assays) and dna E gene product (measured as DNA polymerase). The glycerol gradient fractions contained no detectable rNMP- or dNMP-incorporating activity and no detectable RNase H, DNA-exonuclease, or DNA-endonuclease activity on ØX174 DNA or colicin E_I DNA measured in the presence or absence of ATP. However, glycerol gradient fractions and polyacrylamide gel fractions contained N-ethylmaleimide-resistant DNA-dependent and -independent ribonucleoside triphosphatase activities which appear to be associated with dna B complementing activity.

V. PURIFICATION OF dna C GENE PRODUCT

A. Crude Extract

E. coli HMS-83 (100 g), suspended in 100 ml of 10% sucrose, 0.05 M Tris·HCl, pH 8, and 10^{-3} M dithiothreitol, was lysed with 0.02 M EDTA, 0.1 M KCl, 0.1% Brij 58, 0.2 mg/ml lysozyme at 0°C for 20 min, and then centrifuged at 100,000 *g* for 45 min. The pellet was discarded.

B. Streptomycin Sulfate and Ammonium Sulfate Precipitation

A solution of 20% streptomycin sulfate was added to the crude extract (200 ml) to a final concentration of 4%, and the mixture was centrifuged at 10,000 *g* for 5 min. The supernatant (235 ml) was adjusted to 40% saturation with saturated neutralized ammonium sulfate. After centrifugation at 10,000 *g* for 15 min, the pellet was collected, resuspended in 10 ml of 0.01 M potassium phosphate buffer, pH 7.5, 10^{-3} M dithiothreitol, 5 x 10^{-4} M EDTA, and 20% glycerol (buffer D) and dialyzed against 500 ml of buffer D (changed four times) over 3 hr.

C. DNA Agarose Column Chromatography

The dialyzed ammonium sulfate fraction was diluted to 225 ml with 2×10^{-3} M potassium phosphate buffer, pH 6.8, 10^{-3} M dithiothreitol, 5×10^{-4} M EDTA, and 10% glycerol (buffer E). The final salt concentration was 5×10^{-3} M. The sample was applied to a 100-ml column of denatured calf thymus DNA agarose [27] that was equilibrated with buffer E. The column was washed with 50 ml of buffer E and dna C activity was eluted with 0.05 M Tris·HCl, pH 7.5, 20% glycerol, 10^{-3} M dithiothreitol, 5×10^{-4} M EDTA (buffer F) containing 1 M NaCl. The 1 M salt eluate was adjusted to 50% saturation with solid ammonium sulfate (29.1 g/100 ml). The precipitate was collected by centrifugation, dissolved in 5 ml of buffer F and dialyzed against 1 liter of the same buffer for 3 hr.

D. DEAE-Cellulose Column Chromatography

The DNA agarose fraction (5 ml) was diluted to 10 ml of buffer F and applied to a 35-ml DEAE cellulose column equilibrated with buffer F. The column was washed with 50 ml of buffer F and 5-ml fractions were collected; dna C activity appeared in the effluent that passed directly through the column. Active fractions were pooled (15 ml), and adjusted to 60% saturation with solid ammonium sulfate (36.1 g/100 ml); the precipitate was collected by centrifugation. This material was dissolved in 1 ml of buffer F. The activity, at this stage, was unstable if stored at protein concentrations lower than 0.5 mg/ml. At a protein concentration of 2 mg/ml or higher, dna C activity has been stable over a 2-month period even with repeated freezing (-10°C) and thawing.

E. Glycerol Gradient Centrifugation

A portion of the DEAE-cellulose fraction was dialyzed for 0.5 hr against 0.2 M KCl, 0.02 M potassium phosphate buffer, pH 7.5, 2 x 10^{-3} M dithiothreitol and 5 x 10^{-4} M EDTA. The dialyzed sample (0.2 ml) was layered on a 5-ml linear gradient of 10 to 30% glycerol in the same buffer and centrifuged in the Spinco SW50.1 rotor at 50,000 rpm for 30 hr. The *dna* C activity sedimented through half the gradient. This fraction was unstable, possibly due to its low protein concentration.

The purification, yields, and ratios of *dna* D activity to *dna* C activity are summarized in Table 2. The extent of purification from the ammonium sulfate fraction through the glycerol gradient step was about 175-fold, with a 20% yield.

TABLE 2

Purification of *dna* C Gene Product

Fraction	Total U	Specific activity, U/mg	% recovery	Ratio of *dna* C to *dna* D activity
High-speed supernatant	--	--	--	--
Ammonium sulfate fraction (0-40%)	225	0.4	100	1.1
DNA agarose eluate	160	2.5	71	1.3
DEAE-cellulose eluate	70	27	31	1.6
Glycerol gradient fraction[a]	45	70	20	1.4

[a]This step has been carried out only on a small portion of the DEAE-cellulose eluate. The results presented are calculated for the use of the entire DEAE-cellulose eluate fraction.

F. Properties of dna C (D) Gene Product

The dna C and dna D complementing activities were not separated by the purification procedure (Table 2). Furthermore, all fractions that contained one activity contained both. Within the limits of the assays, dna C and dna D complementing activities sedimented coincidently through glycerol gradients with a molecular weight estimated to be about 25,000. Both activities were N-ethylmaleimide sensitive and both activities were heat inactivated at similar rates. Both dna C and dna D complementing activities were thermolabile in the DEAE-cellulose fraction prepared from the ts mutant when compared with the DEAE-cellulose fraction isolated from the wild type. Since the preparation purified from dna C ts cells was ts in the dna C complementation assay, the material obtained using this assay and purification procedure contains the dna C gene product. Since the activity purified from the dna C ts cells was also ts in the dna D complementation assay, either the product of the dna C gene is also the product of the dna D gene (i.e., dna C and dna D are one locus) or inactivation of the dna C gene product results in the activation of the dna D gene product.

Fractions isolated from dna D ts cells (PC79) were inactive in both dna D and dna C complementation assays. The lack of dna D-complementing activity in fractions prepared from the dna D ts mutant is probably due to the instability of the dna D ts gene product. The lack of dna C complementing activity in the dna D ts preparation again suggests either the products of the two genes are the same or inactivation of one of the gene products results in inactivation of the other. At present, the enzymatic activity that corresponds to the dna C gene product is unknown. The purified preparation did not catalyze incorporation of ribo- or deoxyribonucleotides. It was free of DNA-dependent and -independent ATPase activities. ATP-dependent and -independent DNase and RNase.

The dna C gene product, purified through the DEAE-cellulose step, stimulated ØX174 DNA-dependent dNMP incorporation only in dna C and dna D complementation assays. It was free of dna B and dna G gene products (measured in similar complementation assays) and dna E gene product (measured as DNA polymerase activity using DNase-treated DNA as primer template).

VI. PURIFICATION OF dna G GENE PRODUCT

A. Crude Extract

E. coli HMS-83 (400 g), suspended in 400 ml of 0.02 M potassium phosphate, pH 7.5, 0.05 M KCl, 5 x 10^{-4} M EDTA, 10^{-3} M dithiothreitol, and 10% glycerol, was disrupted by passage through a Manton-Gaulin laboratory homogenizer at 9000-10,000 psi. The crude extract (900 ml) was centrifuged at 100,000 g for 60 min and the pellet discarded.

B. Streptomycin Sulfate and Ammonium Sulfate Precipitation

A solution of 20% streptomycin sulfate was added to the crude extract to a final concentration of 4% and the mixture was centrifuged at 10,000 g for 15 min. The supernatant was adjusted to 40% saturation with solid ammonium sulfate (22.6 g/100 ml); after centrifugation at 10,000 g for 15 min, the supernatant was removed and the pellet was washed successively with 200 ml each of 40%, 30%, and 20% saturated ammonium sulfate solution in 0.02 M potassium phosphate, pH 7.5, 10^{-3} M dithiothreitol, and 5 x 10^{-4} M EDTA. The supernatant obtained after extraction with 20% ammonium sulfate was adjusted to 40% saturation with solid ammonium sulfate (11.3 g/100 ml). The precipitate was collected by centrifugation, dissolved in 100 ml of 0.02 M Tris-HCl, pH 7.5, 10^{-3} M dithiothreitol, 5×10^{-4} M EDTA, and 20% glycerol (buffer G) and dialyzed against buffer G for 6 hr with six 2-liter changes of buffer.

C. DNA Agarose Column Chromatography

The dialyzed ammonium sulfate fraction was diluted to 250 ml with buffer G minus glycerol and applied to a 6 x 40 cm column of denatured calf thymus DNA agarose [27]. It was essential that the salt concentration was 0.005 M or lower before the sample was applied to the column. The column was washed with 500 ml of buffer G containing 10% glycerol and the dna G activity was eluted with buffer G containing 1 M NaCl. The 1 M NaCl eluate was adjusted to 50% saturation with solid ammonium sulfate (29.1 g/100 ml); the precipitate was collected by centrifugation and dissolved in 14 ml of 0.05 M Tris-HCl, pH 7.8, 10^{-3} M dithiothreitol, 5×10^{-4} M EDTA, and 20% glycerol (buffer H).

D. DEAE-Cellulose Column Chromatography

The DNA agarose fraction was dialyzed against 2 liters of buffer H and applied to a DEAE-cellulose column (2.2 x 19 cm) equilibrated with 2 liters of the same buffer. The column was washed with 35 ml of buffer H and then developed with a 600-ml linear gradient from 0 to 0.35 M KCl in buffer H. The dna G activity eluted at 0.1 M KCl.

E. DEAE-Sephadex Column Chromatography

The active fractions (36 ml) were pooled, dialyzed against 2 liters of buffer H, and applied to a DEAE-Sephadex column (2 x 24 cm) equilibrated with the same buffer. The column was washed with 20 ml of buffer H and then developed with a 400-ml linear gradient from 0 to 0.2 M KCl in buffer H. The dna G activity eluted between 0.17 to 0.19 M KCl; these fractions were pooled and adjusted to 50% saturation with solid ammonium sulfate. The pellet obtained after centrifugation at 50,000 g for 20 min was dissolved in 3 ml of buffer H.

F. Glycerol Gradient Centrifugation

A portion of the DEAE-Sephadex fraction was dialyzed against 0.05 M Tris-HCl, pH 7.8, 0.2 M KCl, 10^{-3} M dithiothreitol, 5×10^{-4} M EDTA, and 5% glycerol. The dialyzed sample (0.2 ml) was layered on 6 ml of a 10 to 30% glycerol gradient in the same buffer and centrifuged in the Spinco SW 50.1 rotor at 50,000 rpm for 35 hr. The dna G activity sedimented through two-thirds of the gradient.

TABLE 3

Purification of dna G Gene Product

Fraction	Total protein, mg	Total U	Specific activity, U/mg	% recovery
High-speed supernatant	59,500	--	--	--
20-30% ammonium sulfate fraction	3,240	1375	0.37	100
DNA agarose eluate	100	727	6.7	53
DEAE-cellulose eluate	28.8	650	22.6	47
DEAE-Sephadex eluate	3.8	270	74	20
Glycerol gradient[a]	1.3	270	222	20

[a]This step was carried out with only part of the DEAE-Sephadex eluate. The values reported were calculated assuming that the yield would be the same if the entire fraction were subjected to the glycerol gradient procedure.

The purification and yields of dna G activity carried through the above procedure are summarized in Table 3. The extent of purification from the ammonium sulfate fraction through the glycerol gradient step was about 600-fold with a 20% yield.

G. Properties of dna G Gene Product

The dna G complementing activity purified from dna G ts cells was thermolabile when compared with wild-type dna G activity. In addition, temperature inactivation curves obtained by mixing dna G gene products isolated from the ts mutant and the wild-type were the sum of temperature inactivation curves observed with each preparation. These results support the conclusion that the protein that has been purified is the product of the dna G gene.

The dna G gene product has a molecular weight of about 60,000 (determined by glycerol gradient sedimentation), N-ethylmaleimide is resistant, and binds weakly to single- and double-stranded DNA. No enzymatic activity has been identified as yet that can be ascribed to the dna G protein. The DEAE-Sephadex fraction did not catalyze DNA-dependent incorporation of ribo- or deoxyribonucleotides, nor did it influence RNA or DNA synthesis catalyzed by purified DNA polymerases I, II, or III of E. coli. It was free of detectable DNase and RNase but contained detectable RNase activity, which was N-ethylmaleimide sensitive. Glycerol gradient fractions of dna G gene product contained ATPase and GTPase activities that were stimulated by DNA, but these activities did not sediment coincidentally with dna G activity and were not more thermolabile in preparations from dna G ts cells; thus, these activities may be contaminants. The DEAE-Sephadex fraction was free of dna B and dna C (D) activities (measured by complementation in the ØX174 DNA-dependent assays) and dna E gene product (measured as DNA polymerase activity).

ACKNOWLEDGMENTS

The authors are indebted to Dr. J. Wechsler, Dr. P. Carl, and Dr. Y. Hirota for their generosity in supplying bacterial strains. This work was supported by grants from the National Institutes of Health (GM-13344) and the American Cancer Society (NP-890D). Sue Wickner is a trainee of the National Institute of Health, Ira Berkower is a Medical Scientist Trainee, and Michel Wright is a fellow of the Jane Coffin Childs Memorial Fund for Medical Research.

REFERENCES

1. Y. Hirota, A. Ryter, and F. Jacob, (1968) Cold Spring Harbor Symp. Quant. Biol., 33, 677 (1968).

2. J. A. Wechsler and J. D. Gross, Mol. Gen. Genetics, 113, 273 (1971).

3. P. L. Carl, Mol. Gen. Genetics, 109, 107 (1970).

4. M. L. Gefter, Y. Hirota, T. Kornberg, J. Wechsler, and C. Barnoux, Proc. Natl. Acad. Sci. U.S., 68, 3150 (1971).

5. V. Nusslein, B. Otto, F. Bonhoeffer, and H. Schaller, Nature New Biol., 234, 285 (1971).

6. J. A. Fuchs, H. O. Karlstrom, H. R. Warner, and P. Reichard, Nature New Biol., 238, 69 (1972).

7. R. Moses and C. C. Richardson, Proc. Natl. Acad. Sci. U.S., 67, 674 (1970).

8. H. Schaller, B. Otto, V. Nusslein, J. Huf, R. Herrmann, and F. Bonhoeffer, J. Mol. Biol., 63, 183 (1972).

9. R. B. Wickner and J. Hurwitz, Biochem. Biophys. Res. Commun., 47, 202 (1972).

10. W. T. Wickner, D. Brutlag, R. Schekman, and A. Kornberg, Proc. Natl. Acad. Sci. U.S., 69, 965 (1972).

11. R. Schekman, W. T. Wickner, O. Westergaard, D. Brutlag, K. Geider, L. L. Bertch, and A. Kornberg, Proc. Natl. Acad. Sci. U.S., 69, 2691 (1972).

12. R. B. Wickner, M. Wright, S. Wickner, and J. Hurwitz, Proc. Natl. Acad. Sci. U.S., 69, 3233 (1972).

13. M. Wright, S. Wickner, and J. Hurwitz, Proc. Natl. Acad. Sci. U.S., 70, 3120 (1973).

14. S. Wickner, I. Berkower, M. Wright, and J. Hurwitz, Proc. Natl. Acad. Sci. U.S., 70, 2369 (1973).

15. S. Wickner, M. Wright, and J. Hurwitz, Proc. Natl. Acad. Sci. U.S., 70, 1613 (1973).

16. T. Kornberg and M. L. Gefter, J. Biol. Chem., 247, 5369 (1972).

17. B. Otto, F. Bonhoeffer, and H. Schaller, Europ. J. Biochem., 34, 440 (1973).

18. V. Nusslein, F. Bonhoeffer, A. Klein, and B. Otto, in The Second Annual Harry Steinbock Symposium (R. Wells and R. Inman, eds.), University Park Press, Maryland), 1972.

19. J. L. Campbell, L. Soll, and C. C. Richardson, Proc. Natl. Acad. Sci. U.S., 69, 2090 (1972).

20. J. A. Wechsler, V. Nusslein, B. Otto, A. Klein, F. Bonhoeffer, R. Herrmann, L. Gloger, and H. Schaller, J. Bacteriol., 113, 1381 (1973).

21. R. L. Sinsheimer, in Procedures in Nucleic Acid Research (G. L. Cantoni and D. R. Davies, eds.), Harper and Row, New York, 1966, pp. 659-676.

22. B. Franke and D. S. Ray, Virology, 44, 168 (1970).

23. Bucher, Biochem. Biophys. Acta, 1, 292 (1947).

24. I. Brown, Biochem. Biophys. Acta, 191, 731 (1969).

25. K. Weber and M. Osborn, J. Biol. Chem., 244, 4406 (1969).

26. T. W. Conway and F. Lipmann, Proc. Natl. Acad. Sci. U.S., 52, 1462 (1964).

27. H. Schaller, C. Nusslein, F. J. Bonhoeffer, C. Kurz, and I. Neitzschmann, Europ. J. Biochem., 26, 474 (1972).

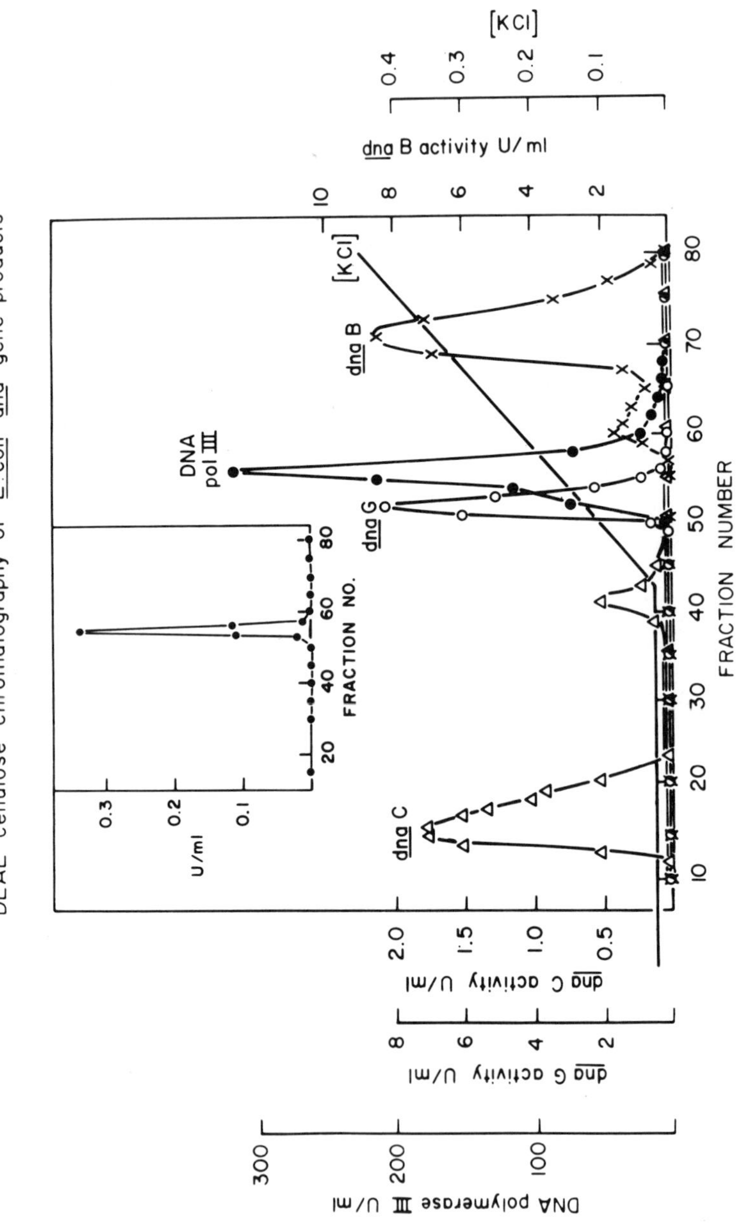
DEAE cellulose chromatography of E. coli dna gene products
DNA pol III
dna G
dna B
dna C
[KCl]
U/ml
FRACTION NO.
FRACTION NUMBER
dna B activity U/ml
dna C activity U/ml
dna G activity U/ml
DNA polymerase III U/ml

FIG. 1. Separation of *E. coli dna* gene products by DEAE-cellulose column chromatograph. Ammonium sulfate I fraction was prepared from 200 g of *E. coli* strain HMS-83 as described in Section II,D.2. This fraction (90 ml; 15.3 mg/ml protein) was dialyzed against 2 liters of 10% glycerol, 1 mM dithiothreitol, 0.5 mM EDTA, and 0.01 M Tris-HCl, pH 7.5, and applied to a 3 x 30 cm DEAE-cellulose column equilibrated with the same buffer. The column was washed with 75 ml of the above buffer and then developed with a 650-ml linear gradient from 0 to 0.6 M KCl in the same buffer; 13-ml fractions were collected. The complementation assays for *dna* B, C, and G gene products were as described in Section II,E; the assay for DNA polymerase III was as previously described [13,14]. The insert figure represents rifampicin-resistant ØX174 DNA-dependent activity that was measured as follows. Each assay (0.05 ml) contained 20 mM Tris-HCl, pH 7.5, 10 mM $MgCl_2$, 4 mM dithiothreitol, 0.04 mM each of dATP, dGTP, dCTP, and [$\alpha^{32}P$] dTTP, 20 μg/ml of rifampicin, 5 mM ATP, 2.5 mM spermidine·HCl, 500 pmole ØX174 DNA and protein fractions. One unit of activity incorporated 1 nmole of dTMP in 30 min at 25°C.

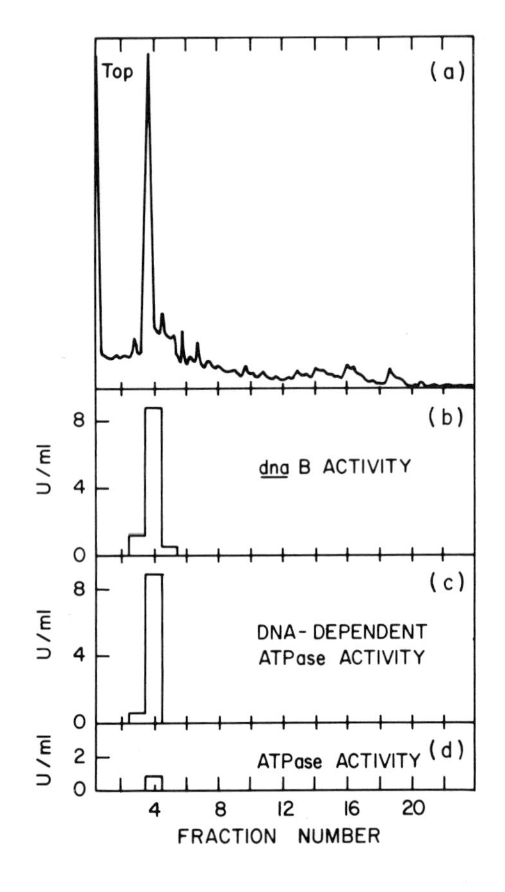

Top
(a)
(b)
8
4
0
U/ml
dna B ACTIVITY
(c)
8
4
0
U/ml
DNA-DEPENDENT
ATPase ACTIVITY
2
0
U/ml
ATPase ACTIVITY
(d)
4
8
12
16
20
FRACTION NUMBER

FIG. 2. Disc gel electrophoresis of dna B gene product. The glycerol gradient fraction was subjected to gel electrophoresis in the system described by Brown [24]. Gels were either stained by the method of Weber and Osborn [25] and traced with a densitometer (A) or sliced with a razor blade into small disks that were incubated 3 hr at 4°C in 10% glycerol, 0.02 M potassium phosphate, pH 7.0, 0.05 M KCl, 1 mM dithiothreitol, 1 mM EDTA, and 0.2 mg/ml bovine serum albumin and assayed for dna B activity (B), DNA-dependent ATPase activity (C), and DNA-independent ATPase activity, (D). ATPase was measured in the following way. Each assay (0.05 ml) contained 10 mM potassium phosphate, pH 7.0, 25 mM KCl, 1 mM $MgCl_2$, 0.5 mg/ml bovine serum albumin, 10 mM N-ethylmaleimide, 1 mM [$\gamma^{32}P$]ATP (10,000 cpm/nmole), and 10 μg/ml rifampicin. Incubations were for 1 hr at 30°C in the presence of 2 nmole of ØX174 single-stranded DNA or in the absence of DNA. Inorganic phosphate was measured by the method of Conway and Lipmann [26]. One unit of ATPase activity catalyzed the production of 1 μmole of inorganic phosphate in 1 hr at 30°C.

Chapter 11B

ISOLATION OF *dna* B-GENE PRODUCT OF *ESCHERICHIA COLI*

Michel Wright and Sue Wickner

Department of Developmental Biology and Cancer
Division of Biological Sciences
Albert Einstein College of Medicine
Bronx, New York

I. INTRODUCTION

The formation of duplex DNA from ØX174 single-stranded DNA by extracts of *Escherichia coli* [1,2] requires the gene product of *dna* B [3]. Inactivated crude extracts prepared from temperature-sensitive *dna* B cells are stimulated by addition of this gene product and provide an assay for the measurement and the purification of the *dna* B gene product [4].

II. BACTERIAL STRAINS

The following strains of *E. coli* K 12 were used: HMS-83 (*pol* A_1, *pol* B_1, *thy*, *lys*) [5] and BT1029 (*pol* A_1, *end*, *thy*, *dna* B ts) [6].

A. Preparation of Fractions for Use in *dna* B Complementation Assay

E. coli strain BT1029, grown to an OD_{590} of 0.55 at 30°C in Hershey broth (5 g NaCl, 5 g Bacto-Peptone, 10 g nutrient broth and 1 g glucose per liter) supplemented with thymine (50 μg/ml) and thiamine (10 μg/ml), was collected by centrifugation at room temperature, resuspended in 0.05 M Tris·HCl, pH 7.5, 10% sucrose (1.5 ml/g wet cells), and frozen in a dry ice-ethanol bath. Frozen cells were kept at -70°C and could be used even after several months of storage.

III. ASSAY OF dna B-GENE PRODUCT

A. Preparation of dna B Receptor Crude Extract

E. coli strain BT1029, grown and frozen as described above, was thawed in a 20°C water bath, adjusted to 0.2 mg/ml egg white lysozyme (Calbiochem), 0.01 M dithiothreitol, and 0.5% Brij 58 and immediately distributed into centrifuge tubes. After 60 min in ice, the mixture was centrifuged for 30 min at 100,000 *g* at 4°C and the supernatant was frozen in 0.5 ml portions; receptor crude extract, when thawed, was stored in ice and used the same day. The *dna* B receptor crude extract (17 mg protein/ml) has been stored frozen for at least 4 months without detectable change in its activity.

B. Preparation of dna B Receptor Ammonium Sulfate Fraction

E. coli BT1029 (300 g of cells), grown and frozen as above, was thawed and adjusted to 0.2 mg/ml lysozyme; after mixing, the suspension was immediately distributed into centrifuge tubes and incubated for 60 min at 0°C, then for 3 min at 20°C. The lysate was centrifuged for 30 min at 100,000 *g* and the supernatant (340 ml, 10 mg protein/ml) was frozen overnight. After thawing at 20°C, the fraction was adjusted to 60% with solid ammonium sulfate (36.1 g/100 ml), kept 10 min in ice, and centrifuged 30 min at 100,000 *g*. The pellet was resuspended successively in 110 ml of 40%, 50 ml of 30%, and 20 ml of 20% saturated ammonium sulfate solution containing 0.05 M morpholinopropane sulfonate (MOPS), pH 7.5, 0.1 mM EDTA, 1 mM dithiothreitol (buffer A). After each resuspension, the precipitate was centrifuged 30 min at 100,000 *g*. The supernatant obtained after centrifugation of the 20% saturated ammonium sulfate solution was adjusted to 50% saturation with solid ammonium sulfate (17.5 g/100 ml) and centrifuged 30 min at 100,000 _g_. The pellet was dissolved in 4.6 ml of 0.05 M MOPS, pH 7.5, 0.5 mM EDTA, 1 mM dithiothreitol, 0.01 M $MgCl_2$, 1 mM ATP (buffer B), and dialyzed 2 hr against 250 ml and overnight against 500 ml of the same buffer. The ammonium sulfate fraction (12 mg of protein/ml) was inactivated by heating for 12 min at 41°C and insoluble material was removed by centrifugation for 15 min at 100,000 _g_. The _dna_ B receptor ammonium sulfate fraction was stable to repeated freezing and thawing.

C. Complementation Assay for _dna_ B

Each assay (0.1 ml) contained 2.5 μg of rifampicin,* 1 μmole of MOPS pH 7.5, 20 nmoles of dithiothreitol, 10 nmoles of EDTA, 1 μmole of $MgCl_2$, 400 nmoles of ATP, 0.6 nmole of ØX174 single-stranded DNA, 1.25 nmoles of [$\alpha^{32}P$]dTTP or dATP (50-2000 cpm/pmole), 2.5 nmoles of each of the three other dNTPs, 0.04 ml of _dna_ B receptor crude extract, 0.01 ml of _dna_ B receptor ammonium sulfate fraction, and protein fractions as indicated. The reaction mixture was incubated for 30 min at 30°C and acid-insoluble radioactivity determined. One unit (U) of _dna_ B complementing activity incorporated 1 nmole of dTMP or dAMP under the above conditions. Specific activity is expressed in units per milligram of protein. Under the conditions of the complementation assay, the rate of dAMP incorporation was linear between 10 and 40 min at 30°C and the rate of dAMP incorporation was directly proportional to the amount of _dna_ B gene product up to 0.035 U. The incorporation of dAMP and dTTP was not dependent on dNTPs, presumably due to their presence in the receptor crude extracts. Both _dna_ B re-

---

*Rifampicin was without effect on the ØX system. However, in order to avoid any role of DNA-dependent RNA polymerase in this reaction, this antibiotic was added to all reaction mixtures.

ceptor crude extract and dna B receptor ammonium sulfate fractions were required in the complementation assay of dna B gene product. The omission of either one resulted in no detectable activity. However, the dna B product from streptomycin sulfate and ammonium sulfate fractions was not completely dependent upon the addition of the dna B receptor ammonium sulfate fraction.

IV. PURIFICATION OF dna B-GENE PRODUCT

A. Streptomycin Sulfate Fraction

Frozen cells (400 g, E. coli strain HMS-83) were suspended in 400 ml of 0.05 M MOPS, pH 7.5, 0.5 mM EDTA, 1 mM dithiothreitol, and 0.05 M KCl and disrupted by passage through the Manton-Gaulin laboratory homogenizer at 650 kg/cm^2. The crude extract** (815 ml) was adjusted to a final concentration of 4% with freshly prepared 20% streptomycin sulfate, mixed, and centrifuged at 100,000 g for 30 min. The supernatant (785 ml; 16.4 mg protein/ml) could be stored frozen several days without loss of dna B complementing activity.

B. Ammonium Sulfate Fraction

The streptomycin sulfate fraction was adjusted to 40% with solid ammonium sulfate (22.6 g/100 ml). After 10 min the mixture was centrifuged 30 min at 100,000 g. The pellet was then successively resuspended in 200, 115, and 90 ml of buffer A containing 40, 30, and 20% saturated ammonium sulfate, respectively.

**Activity present in crude extracts could not be measured because dNMP incorporation was independent of added ØX DNA.

After each resuspension, the mixture was centrifuged 30 min at 100,000 g. The supernatant obtained after extraction with the 20% saturated ammonium sulfate solution was precipitated by addition of ammonium sulfate to 50% saturation (17.5 g/100 ml) and centrifuged 30 min at 100,000 g. The pellet was dissolved in 30 ml of buffer B and dialyzed 2 hr against two 1-liter changes and overnight against 1 liter of the same buffer. The ammonium sulfate fraction (34 ml; 18 mg protein/ml) could be stored frozen several months without loss of dna B complementing activity.

C. DEAE-Cellulose Fraction I

The dialyzed ammonium sulfate fraction, diluted fivefold with 0.05 M MOPS, pH 7.5, 0.5 mM EDTA, 1 mM dithiothreitol, and 20% glycerol (buffer C), was applied to a DEAE-cellulose column (25 x 4.5 cm) equilibrated with the same buffer. The column was washed successively with 300 ml of buffer C, 300 ml of buffer C containing 0.1 M NaCl, and 1000 ml of buffer C containing 0.2 M NaCl. The dna B complementing activity was eluted with 750 ml of buffer C containing 0.3 M NaCl; 20-ml fractions were collected, and fractions containing the dna B complementing activity were pooled, adjusted to 60% saturation with ammonium sulfate (36.1 g/100 ml), and centrifuged for 30 min at 100,000 g. The pellet was dissolved in 4 ml of buffer B containing 20% glycerol and dialyzed for 180 min against 250 ml of the same buffer (changed six times). The DEAE-cellulose fraction I (5.9 ml; 0.90 mg protein/ml) could be stored frozen overnight.

D. DEAE-Cellulose Fraction II

The dialyzed DEAE-cellulose fraction I was diluted fourfold with 0.05 M Tris, pH 8.5, 0.1 mM EDTA, 1 mM dithiothreitol, 20% glycerol, and applied to a DEAE-cellulose column (8 x 1.2 cm) equilibrated with the same buffer. The column was washed with 20 ml of the same buffer and then subjected to a 140-ml linear gradient between 0 and 0.5 M NaCl in the same buffer. The dna B complementing activity eluted at a salt concentration between 0.25 and 0.3 M NaCl. All fractions containing dna B activity were pooled, vacuum concentrated, and dialyzed in collodion bags (Schleicher and Schuell, Inc.) against buffer B containing 20% glycerol. The fluid was changed each hour over a 12-hr period. The DEAE-cellulose fraction II (2 ml; 1.07 mg protein/ml) was stored frozen; 60% of the activity remained after 1 month of storage.

E. Glycerol Gradient Fraction

A portion of the DEAE-Cellulose fraction II (0.2 ml) was diluted to 0.3 ml with buffer B, applied to a 5-ml 15-35% glycerol gradient in the same buffer, and centrifuged 14 hr at 48,400 rpm

TABLE 1

Purification of dna B Gene Product

Fraction	Total U	Specific activity, U/mg	% recovery
Streptomycin sulfate	3700	0.3	100
Ammonium sulfate	1677	2.8	45
DEAE-cellulose I	366	69	9.9
DEAE-cellulose II	436	204	11.8
Glycerol gradient[a]	192	1200	5.2

[a]This step was performed with only part of the DEAE-cellulose II fraction. The values reported assume that the yield and purification would be the same if the entire fraction were subjected to the glycerol gradient procedure.

at 4°C (SW 65 Spinco rotor). dna B activity sedimented through half of the gradient. Fractions containing dna B complementing activity (0.8 ml; 0.02 mg protein/ml) were stored at -10°C. The dna B complementing activity was purified about 4000-fold with a 5% overall yield (Table 1). The final preparation retained 35% of its activity after 5 days.

V. PROPERTIES OF PURIFIED dna B GENE PRODUCT

The dna B gene product has a molecular weight of between 2.5 and 3.0 x 10^{-5}, and its complementing activity was unaffected by treatment with N-ethylmaleimide. The dna B gene product (DEAE-cellulose fraction II) was free of dna C (D), E, and G gene products measured by the ØX DNA complementation assays. The glycerol gradient fraction was examined for other enzymatic activities. No rNMP or dNMP incorporating activity was detected; there was no detectable RNase H, DNA-exonuclease, or DNA-endonuclease activity on ØX174 DNA or colicin E_I DNA measured in the presence or absence of ATP. However, dna B gene product (glycerol fractions) contained an N-ethylmaleimide-resistant DNA-stimulated ribotriphosphatase activity that appears to be associated with dna B complementing activity.

A slight modification of the procedure described here and in Ref. 4 has resulted in essentially homogeneous preparations of the dna B gene product (see Chapter 11A and Ref. 7).

ACKNOWLEDGMENTS

The authors are indebted to Mrs. B. Philips for her skillful technical assistance. This work was supported by grants from the National Institutes of Health (GM-13344) and the American Cancer Society (NF-890D). Michel Wright is a fellow of the Jane Coffin Childs Memorial Fund for Medical Research and Sue Wickner is a trainee of the National Institutes of Health.

REFERENCES

1. W. T. Wickner, D. Brutlag, R. Schekman, and A. Kornberg, Proc. Natl. Acad. Sci. U.S., 69, 965 (1972).

2. R. B. Wickner, M. Wright, S. Wickner, and J. Hurwitz, Proc. Natl. Acad. Sci. U.S., 69, 3233 (1972).

3. R. Schekman, W. T. Wickner, O. Westergaard, D. Brutlag, K. Geider, L. L. Bertch, and A. Kornberg, Proc. Natl. Acad. Sci. U.S., 69, 2691 (1972).

4. M. Wright, S. Wickner, and J. Hurwitz, Proc. Natl. Acad. Sci. U.S., 70, 3120 (1973).

5. J. L. Campbell, L. Soll, and C. C. Richardson, Proc. Natl. Acad. Sci. U.S., 69, 2090 (1972).

6. J. A. Wechsler, V. Nusslein, B. Otto, A. Klein, F. Bonhoeffer, R. Herrmann, L. Gloger, and H. Schaller, J. Bacteriol., 113, 1381 (1973).

7. S. Wickner, M. Wright, and J. Hurwitz, Proc. Natl. Acad. Sci. U.S., 71, 783 (1974).

Chapter 12

A DNA-CELLULOSE PROCEDURE FOR THE PURIFICATION OF THE *ESCHERICHIA COLI* PROTEIN ω

Jonathan O. Carlson and James C. Wang

Department of Chemistry
University of California
Berkeley, California

I. INTRODUCTION

The *E. coli* protein ω causes the reduction in superhelical density of negatively twisted superhelical DNA. This is apparently accomplished by the introduction of a transient swivel into the DNA molecule [1-3]. The role of ω as a swivelase in vitro has raised the possibility that it might participate in biological processes requiring a swivel, such as replication. The purification procedure presented here allows a more rapid preparation of the protein and yields a product of higher purity than the previously published procedure [1,4].

II. ASSAY OF THE ω PROTEIN

There are several assays for the ω protein including band sedimentation in a ethidium-containing medium [1], density gradient equilibrium centrifugation in ethidium-CsCl [2,3], and electron microscopy [1]. Only the assay using band sedimentation is described here. Procedures for the other methods have been described elsewhere [4].

The band sedimentation assay is based on the fact that the sedimentation coefficient of a covalently closed DNA is dependent on its superhelical density. Negative superhelical DNA can be obtained either from natural sources such as intracellular covalently closed λ DNA [5], ØX174 RFI [5], or phage PM2 [6] or by covalent closure of nicked circular DNA in vitro by ligase in the presence of ethidium [7,8]. The DNA of the phage PM2 grown on *Pseudomonas* BAL31 is highly twisted when isolated from the phage particle, and it can be obtained in large amounts with relative ease. When extracted from freshly purified phage, over 95% of the DNA is in the twisted form and can be used directly in the assays. The DNA can be stored for long periods in 2 M NaCl,

0.01 M EDTA, pH 8. For ω assays the DNA is dialyzed into the reaction buffer, 2 mM $MgCl_2$, 1 mM EDTA, 10 mM Tris·HCl, pH 8. In this medium the DNA is stable for at least a month when free of chemical and biologic contamination.

Dilutions of ω can be made with a solution of 1 mg/ml bovine serum albumin (BSA) in the reaction buffer. Before use the BSA should be checked for nuclease activity on superhelical DNA. Incubation of the BSA solution with a sample of superhelical DNA followed by band sedimentation in 0.1 M KOH, 3 M CsCl, 0.01 M EDTA provides a sensitive test for nicking activity.

A. Procedure

The reaction mixture is made by mixing 44 μl of reaction buffer, containing 0.5 to 0.8 μg of DNA, with 3 μl of the BSA solution described above, and 3 μl of the ω solution. The reaction mixture is incubated at 30°C for 15 min. The reaction is terminated by the addition of 20 μl of 4 M NaCl, 0.02 M EDTA to the reaction mixture.

Thirty-five microliters of the reaction mixture is placed in the sample well of a double sector 30-mm band-forming centerpiece [9,10]. Thirty-five microliters of 2 M NaCl, 0.01 M Na_3EDTA is placed in the reference well. The bulk sedimenting medium is 3 M CsCl, 0.01 M Na_3EDTA with the proper amount of ethidium bromide added. The ethidium bromide concentration is chosen such that the sedimentation coefficient of the unreacted superhelical DNA is close to its minimum value. Reduction of the superhelical density by ω then results in an increase in the sedimentation coefficient of the DNA. For PM2 DNA, 6 μg/ml of ethidium is adequate. At this concentration unreacted DNA sediments at 12S and the reacted DNA sediments at 15 to 16S. The reacted samples are centrifuged at the appropriate speed depending on the size of the DNA (31,410 rpm is convenient for PM2) for 2 to 3 hr in an analytical centrifuge equipped with a photoelectric scanning system (Spinco). The double-sector optics are used to cancel out the absorption of ethidium in the sedimenting medium. The sedimentation is monitored with 265 nm light.

The assay can be made semiquantitative if need be by following the relative amounts of slow- and fast-sedimenting species with dilution of the ω solution.

III. PURIFICATION PROCEDURE

Since nuclease activity interfers with the ω assay, it is more convenient to isolate the ω protein from endonuclease I-deficient cells such as *E. coli* 1100 [11]. The entire procedure is performed at 4°C.

A. Cell Disruption

Bacteria are suspended in 0.1 M glycylglycine pH 7.0 and disrupted by sonication or by three or four passes through a Manton-Gaulin press [12] or by grinding with glass beads in a Waring Blendor. After the disruption the suspension is centrifuged at 20,000 *g* for 30 min. The protein concentration in the supernatant is usually between 20 and 40 mg/ml.

B. Streptomycin Sulfate Precipitation

The supernatant is adjusted to a protein concentration of 19 mg/ml with 0.1 M glycylglycine pH 7.0. To this solution 0.2 vol of 5% streptomycin sulfate is added dropwise with stirring. After stirring for ½ hr, the suspension is centrifuged at 15,000 *g* for

15 min. The supernatant is diluted with 2 vol of cold distilled water, and 1 vol of 5% streptomycin sulfate is added dropwise with stirring. After stirring for ½ hr, the suspension is again centrifuged at 15,000 g for 15 min.

C. Ammonium Sulfate Precipitation

The streptomycin sulfate supernatant is made 20% w/w in $(NH_4)_2SO_4$ by adding 0.25 g of solid enzyme grade $(NH_4)_2SO_4$ per gram of solution. The $(NH_4)_2SO_4$ is added slowly with constant stirring. After the $(NH_4)_2SO_4$ has dissolved, the stirring is continued for ½ hr and the suspension is centrifuged at 15,000 g for 15 min. To the supernatant an additional 0.14 g $(NH_4)_2SO_4$ per gram of supernatant is slowly added with stirring to raise the $(NH_4)_2SO_4$ concentration to 28% w/w. After the $(NH_4)_2SO_4$ has dissolved, the stirring is continued for ½ hr and the suspension is centrifuged at 15,000 g for 15 min. The precipitate is dissolved in 0.25 M NaCl, 1 mM EDTA, 1 mM-mercaptoethanol, 20 mM Tris·HCl, pH 7.5, and dialyzed for several hours against the same buffer. About 1 g of protein is present in this fraction starting from 100 g of frozen bacteria.

D. DNA-Cellulose Chromatography

DNA-cellulose is prepared by the method of Alberts and Herrick [13] except that the lyophilized DNA-cellulose is suspended in ethanol (200 ml/100 g cellulose) and irradiated with a Mineralight low-pressure mercury lamp at a distance of 30 cm for a total of 60 min with stirring as in the method of Litman [14]. The DNA-cellulose contains about 400 μg of native calf thymus DNA per packed milliliter of column material as determined by reaction with pancreatic DNase.

To the dialyzed $(NH_4)_2SO_4$ fraction is added 0.1 vol of glycerol and the fraction is applied to a DNA-cellulose column (2.5 cm x 21 cm) previously equilibrated with 0.25 M NaCl, 20 mM Tris·HCl, pH 7.5, 1 mM EDTA, 1 mM mercaptoethanol, 10% glycerol, and 0.1 mg/ml BSA. After loading, the column is washed with this buffer until the A_{280} of the effluent drops to the baseline absorbance. The ω activity is eluted from the column by increasing the NaCl concentration in the column buffer (20 mM Tris·HCl, pH 7.5, 1 mM EDTA, 1 mM mercaptoethanol, 10% glycerol, 0.1 mg/ml BSA). This can be done by applying either a linear gradient or a step gradient. The ω activity is eluted between 0.55 and 1 M NaCl.

E. Comments on the Purification

The active ω fractions from the DNA-cellulose column may be checked for contamination with other protein species by SDS-polyacrylamide gel electrophoresis [15]. The monomer and dimer forms of the BSA present in the buffer provide convenient internal molecular weight standards. The ω protein has a molecular weight of 100,000 to 110,000 by this method. The early fractions from a linear NaCl gradient may show contaminating bands. However, the later fractions should be apparently free of contamination. The contaminating proteins can be eliminated by rechromatography of the active fractions on a small DNA-cellulose column (1.6 cm x 4.5 cm).

The ω protein prepared by this procedure appears to be more than 90% pure, aside from the BSA present in the elution buffer, as judged by the absence of other protein bands after SDS-polyacrylamide gel electrophoresis. The ω protein shows no DNA polymerase activity; DNA polymerase also has a molecular weight of about 109,000 [16]. The overall yield is about 0.1 mg of protein from 100 g of cells, but the percentage of the theoretical yield is not known since the activity cannot be detected until the ammonium sulfate precipitation step due to interfering DNA or streptomycin earlier in the procedure.

REFERENCES

1. J. C. Wang, *J. Mol. Biol.*, *55*, 523 (1971).
2. J. C. Wang, in *DNA Synthesis in vitro* (R. B. Inman and R. D. Wells, eds.), University Park, Baltimore, 1973, p. 163.
3. J. J. Champoux and R. Dulbecco, *Proc. Natl. Acad. Sci. U.S.*, *69*, 143 (1972).
4. J. C. Wang, in *Methods in Enzymology* (L. Grossman and K. Moldave, eds.), Vol. 29, Academic, New York, 1974, p. 197.
5. J. C. Wang, *J. Mol. Biol.*, *43*, 263 (1969).
6. R. T. Espejo, E. S. Canelo, and R. L. Sinsheimer, *Proc. Natl. Acad. Sci. U.S.*, *63*, 1164 (1969).
7. J. C. Wang, *J. Mol. Biol.*, *43*, 25 (1969).
8. J. C. Wang, in *Procedures in Nucleic Acid Research* (G. L. Cantoni and D. R. Davies, eds.), Vol. 2, Harper and Row, New York, p. 347.
9. J. Vinograd, R. Bruner, R. Kent, and J. Weigle, *J. Mol. Biol.*, *49*, 902 (1963).
10. J. Vinograd, R. Radloff, and R. Bruner, *Biopolymers*, *3*, 481 (1965).
11. I. R. Lehman, G. G. Roussos, and E. A. Pratt, *J. Biol. Chem.*, *237*, 819 (1962).
12. S. E. Charm and C. C. Matteo, in *Methods in Enzymology* (W. Jakoby, ed.), Vol. 22, Academic, New York, 1971, p. 483.
13. B. Alberts and G. Herrick, in *Methods in Enzymology* (L. Grossman and K. Moldave, eds.), Vol. 21, Academic, New York, 1971, p. 198.

14. R. M. Litman, J. Biol. Chem., 243, 6222 (1968).

15. O. Gabriel, in Methods in Enzymology (W. Jakoby, ed.), Vol. 22, Academic, New York, 1971, p. 565.

16. T. M. Jovin, P. T. Englund, and L. L. Bertsch, J. Biol. Chem., 244, 2996 (1969).

Chapter 13

DNA POLYMERASE FROM WILD-TYPE AND MUTANT T4 BACTERIOPHAGE

Nancy G. Nossal

Laboratory of Biochemical Pharmacology
National Institute of Arthritis, Metabolism,
and Digestive Diseases
National Institutes of Health
Bethesda, Maryland

I. INTRODUCTION

The DNA polymerase determined by the phage gene 43 consists of a single polypeptide of molecular weight 110,000 [1,2]. The enzyme has both polymerase and 3' to 5' exonuclease activities.

The following method for the purification of the wild-type T4 DNA polymerase yields a homogeneous preparation of the enzyme after three chromatographic steps. Modifications of this method are described for the purification of the B22 nuclease, an amber peptide (molecular weight 80,000) produced by the gene 43 mutant _am_ B22 which contains the nuclease but not the polymerase activity of the wild-type enzyme [3], as well as for the purification of the polymerases from temperature-sensitive mutants in this gene. Finally, a method for obtaining the DNA polymerase as a side product of the purification of the gene 32 DNA unwinding protein [4] is outlined.

II. ASSAYS

A. Polymerase Activity

Polymerase activity is measured under the conditions described by Goulian, Lucas, and Kornberg [1] in a reaction mixture containing 67 mM Tris-HCl (pH 8.8), 6.7 mM $MgCl_2$, 16.7 mM ammonium sulfate, 6.7 μM EDTA, 10 mM mercaptoethanol, bovine serum albumin (0.2 mg/ml), 10 nmoles of salmon sperm DNA denatured with alkali (expressed as nucleotide equivalents) [5], and 33 μM in each of the four deoxynucleoside triphosphates (one labeled) in a total volume of 50 μl. Polymerase assays of crude fractions also contained 0.33 mM KF to inhibit the T4-induced dCTPase [1]. The enzyme is diluted in a solution containing 25% glycerol, 1 mg/ml bovine serum albumin, and 10 mM Tris-HCl (pH 8.8). Although

T4 DNA contains hydroxymethyl cytosine, dCTP can be substituted for hydroxymethyl dCTP in vitro [1]. Radioactivity in TCA-insoluble polymer is determined by a modification of the method of Furano [6]. A glass fiber filter (Whatman GF/C) of 12.5 cm diameter was ruled with a number 1 pencil into a grid containing 16 squares of 2.2 cm^2. The filter was wetted with 10% TCA and air dried. An aliquot of each reaction mixture (40 μl) was pipeted onto a square. After all the samples had been taken, the filter was washed successively on a Büchner funnel with 50 ml cold 10% trichloroacetic acid, three times with 50 ml ice cold distilled water, and once with 20 ml ethanol; dried briefly (10 min) under a heat lamp, cut into squares, and counted in a toluene-based scintillation solution [7]. A unit is defined as the amount of enzyme that will catalyze the incorporation of 10 nmoles of total nucleotide in 30 min at 37°C.

Although denatured salmon sperm DNA has been used to monitor the enzyme purification, different batches of this DNA vary markedly in their efficiency as template primers. Polymerase activity with denatured T7 DNA measured under the same conditions is more reproducible from preparation to preparation and therefore is a more reliable template to use to compare the polymerase activity of different mutants.

B. Exonuclease Activity

Exonuclease activity is measured in the same incubation mixture used to assay the polymerase except that the volume is increased to 0.15 ml, 100 nmoles of heat-denatured *E. coli* [^{3}H]DNA [8] is the substrate, and deoxynucleoside triphosphates are omitted. The reaction is terminated by the addition of sperm DNA (0.1 ml of 2.5 mg/ml) and 0.25 ml of 0.5 M cold perchloric acid. After centrifugation, 0.2 ml of the supernatant fraction is counted in a Triton-toluene scintillation solution [9]. A unit of exonuclease catalyzes the hydrolysis of 10 nmoles of nucleotide in 30 min at 37°C.

A spectrophotometric assay can also be used to conveniently monitor the exonuclease activity during purification [3]. The substrate is a mixture of ethanol-precipitable oligonucleotides that is prepared from a pancreatic DNase digest of sperm DNA (100% soluble in trichloroacetic acid) using the conditions for digestion and precipitation described by Oleson and Koerner [10]. The hyperchromic shift at 260 nm due to exonuclease digestion is followed at 37°C in a mixture containing 50 mM Tris-HCl (pH 7.5), 6.3 mM $MgCl_2$, and 1.1 absorbance units of the digested sperm DNA. A unit is defined as an increase in absorbance of 1 optical density unit per minute at 37°C.

III. INFECTED CELLS

Phage lysates were prepared and titered by standard methods [11]. It is convenient to prepare the enzyme from cells infected with phage containing an amber mutation in a gene for another early function since cells infected with such mutants do not lyse and continue to make many early enzymes much longer than cells infected with wild-type phage [12]. Bacteria infected with T4 *am* N82 (gene 44) were used to purify the wild-type enzyme. Polymerase induced by temperature-sensitive mutants was prepared from cells infected with phage containing the *am* N82 (gene 44) mutation as well as a temperature-sensitive mutation in the polymerase gene (gene 43). T4 *am* B22 (gene 43) without any additional mutation was used to prepare the B22 nuclease. *E. coli* CR63 [13] was the amber permissive host, and *E. coli* ER21, an endonuclease I-deficient derivative of E. coli B [14], was the nonpermissive host. Double mutants were constructed by crossing the parents, each at a multiplicity of 5 in E. coli CR63 at 30°C. The isolated mutants failed to complement either parent in spot tests [13] and grew on E. coli CR63 at 30°C but not 43°C. E. coli ER21 was obtained from Dr. J. Eigner (Washington University School of Medicine), E. coli CR63 from Dr. R. Edgar (University of California at Santa Cruz), T4 am N82 from Dr. J. Wiberg (University of Rochester), and T4 gene 43 mutants from Dr. J. Drake (University of Illinois).

E. coli ER21 was grown in a glycerol-salts medium [15] in a 50- or 300- liter fermentor to a concentration of about 1.6×10^9 per ml, which under these conditions of very vigorous aeration is mid-log phase, and infected at a multiplicity of 8 with the appropriate mutant. This medium contains per liter: Na_2HPO_4, 10.5 g; KH_2PO_4, 4.5 g; NH_4Cl, 1 g; $MgSO_4$, 0.3 g; casaminoacids (Difco, technical grade), 15 g; glycerol, 30 g; 1 M $CaCl_2$, 0.3 ml; and 1% gelatin, 1 ml. Sixty minutes after the infection, the culture was cooled by circulating cold water around the fermentor, and the cells were harvested in a Sharples centrifuge. More than 99% of the cells were killed by the infection. The cells were grown and infected at 37°C for amber mutants and at 30°C for temperature-sensitive mutants. Approximately 1 kg of infected cells was obtained per 300 liters. Cell pastes were stored at -70°C.

Smaller batches can be prepared by infecting cells grown to 5×10^8/ml using forced air or rotary shaking, and cooling the infected culture by the addition of 2 vol of ice before centrifugation [3].

IV. ENZYME PURIFICATIONS

A. Wild-Type T4 DNA Polymerase

The following procedure was used to purify the wild-type enzyme from 500 g of *E. coli* ER21 infected with *am* N82 [2]. The procedure can be scaled down by adjusting the size of columns and gradients proportionally. All steps were carried out at 0° to 4°C.

Frozen cells (500 g) were suspended in 1250 ml of extraction buffer containing 50 mM Tris-HCl (pH 7.5), 10 mM mercaptoethanol, 1 mM EDTA, and 5.6 mM phenylmethylsulfonyl fluoride (PMSF). (The latter was added as a precautionary measure since it is an inhibitor of some proteolytic enzymes [16]. However, there is no evidence to date that its presence during purification alters the characteristics of the final polymerase preparation.) The cells were broken by passage through a Gaulin homogenizer press two times at 9000 psi. The extract was rapidly chilled to 4°C in an ice bath after each passage. Smaller batches of cells have been broken by the French pressure cell or by grinding with alumina [3]. The crude extract was centrifuged for 60 min at 15,000 *g*. The supernatant (1600 ml) was then centrifuged for 4 hr at 100,000 *g* in the Spinco No. 30 rotor. This required four centrifugations since the rotor capacity is about 360 ml. The 100,000 *g* supernatant was diluted to 2800 ml with the extraction buffer to lower the conductivity of the fraction to 0.85 mmho, which is equivalent to that of 0.045 M KCl in 25% glycerol. This dilution is essential to ensure binding of the polymerase to DEAE-cellulose.

1. DEAE-Cellulose Chromatography

The diluted fraction was applied to a DEAE-cellulose column (Whatman DE23, 8.5 x 60 cm) that had been equilibrated with 50 mM

Tris-HCl (pH 7.5) and 25% glycerol. The column was washed with 175 ml of the same buffer containing in addition 10 mM mercaptoethanol and 5.6 mM PMSF (buffer A). The protein was eluted by a linear gradient formed from 800 ml buffer A and 800 ml of this buffer containing 50 mM KCl, followed by a second linear gradient formed from 4 liters of buffer A with 50 mM KCl and 4 liters of buffer A containing 0.15 M KCl. Twenty milliliter fractions were collected into tubes containing 2 ml of a solution of 50 mM Tris-HCl (pH 7.5), 25% glycerol, and 56 mM PMSF, since PMSF is retained on the column. PMSF is not soluble at this concentration, but dissolves after tenfold dilution by the column fractions. The flow rate was approximately 250 ml/hr. The T4 DNA polymerase is eluted in a sharp peak between 60 and 80 mM KCl. A much smaller peak of polymerase, which is present in uninfected cells, is eluted at 0.1 to 0.12 M KCl.

2. Hydroxyapatite Chromatography

Eight hundred milliliters (75%) of the pooled polymerase fractions from the DEAE-cellulose column was applied directly to a hydroxyapatite column (2.8 x 11.5 cm) that was prepared as described by Levin [17] and equilibrated with 20 mM Tris-HCl (pH 7.5), 0.1 M KCl, and 25% glycerol (buffer B). The column was washed with 50 ml of buffer B containing 10 mM mercaptoethanol and 5.6 mM PMSF. The enzyme was eluted with a linear gradient formed from 500 ml of this buffer and 500 ml of the same buffer with 10% saturated ammonium sulfate. Ammonium sulfate solutions were prepared by appropriate dilution of a saturated solution that had been stored at 4°C. Thus, 10% saturated refers to a tenfold dilution of the saturated solution. Five milliliter fractions were collected. The initial flow rate of 60 ml/hr decreased during the chromatography. The polymerase was eluted as a sharp peak between two larger peaks of protein at 4% saturation of ammonium sulfate.

3. DNA Cellulose Chromatography

The pooled hydroxyapatite fraction was dialyzed for 5 hr against three changes of 1 liter of 50 mM Tris-HCl (pH 7.5), 50 mM KCl, 10 mM mercaptoethanol, 1 mM EDTA, and 25% glycerol (buffer C). The dialyzed fraction (800 ml) was applied to a column of DNA cellulose (2 x 14 cm) that had been prepared as described by Alberts and Herrick [18] and equilibrated with buffer C. After loading, the column was washed with 60 ml of buffer C, and the enzyme was eluted with a linear gradient formed from 200 ml buffer C and 200 ml of buffer C containing 0.6 M KCl. Two milliliter fractions were collected at 60 ml/hr. The enzyme eluted as a single peak at 0.2 to 0.23 M KCl. The specific activities for both the polymerase and exonuclease activities were constant throughout the peak.

4. Concentration on Phosphocellulose

The pooled fractions were diluted with an equal volume of 50 mM Tris-HCl (pH 7.5), 25% glycerol, and 10 mM mercaptoethanol (buffer D) and loaded immediately onto a phosphocellulose column (Whatman P-11) that had been equilibrated with buffer B containing 0.1 M KCl; the polymerase eluted with buffer B containing 0.4 M KCl. The enzyme was recovered in a total volume of 4 ml, and the pooled fraction was brought to a final concentration of 50% glycerol, 50 mM Tris-HCl (pH 7.5), 10 mM mercaptoethanol, and 0.25 M KCl. The enzyme was free of contaminating DNA endonucleases, using λ DNA as the substrate, and gave a single band on disc gel electrophoresis in the presence of SDS and mercaptoethanol [2]. The enzyme has been stored at -70°C for three years with a 25% loss in activity.

TABLE 1

Purification of Wild-Type T4 DNA Polymerase[a]

Fraction	Polymerase activity,[b] units/mg	Nuclease activity,[c] units/mg	Protein concentration, mg/ml	Total protein, mg
I. 100,000 *g* extract	104	1.3	3.50	9,700
II. DEAE-cellulose	2,630	13.4	0.16	190
III. Hydroxyapatite	19,700	56.9	0.25	10.5
IV. DNA-cellulose after concentration	32,000	140.0	0.64	3.8

[a]The cell paste was 500 g. Seventy-five percent of fraction II was carried through subsequent steps.

[b]Denatured salmon sperm DNA was the template primer.

[c]Nuclease was assayed by the spectrophotometric procedure.

The purification of the wild-type enzyme is summarized in Table 1. The best results have been obtained when the purification has been carried out as rapidly as possible. In view of our subsequent experience with the mutant enzymes, it seems likely that phosphocellulose chromatography under conditions described for the temperature-sensitive mutant enzymes could be used in place of DNA-cellulose chromatography for the wild-type enzyme.

The wild-type polymerase can also be obtained as a byproduct of the purification of the gene 32 DNA unwinding protein. The polymerase is found in the fraction obtained by stepwise elution of the DNA cellulose column with 0.15 and 0.4 M NaCl using the 32 protein purification procedure of Alberts and Frey [4], with polyethylene glycol used to remove the nucleic acids as described by Alberts and Herrick [18]. The pooled polymerase fractions were applied directly to a hydroxyapatite column, eluted as described in the above procedure for the polymerase, and then purified by phosphocellulose chromatography as described below for the temperature-sensitive mutants. The product of this purification scheme appears to be indistinguishable from that obtained by the first procedure. This procedure is recommended only if the 32 protein is also desired since it is more laborious than the direct purification described for the polymerase.

B. T4 DNA Polymerase from Temperature-Sensitive Mutants of T4 Bacteriophage

The DNA polymerase induced by T4 *ts* L88 [19] and T4 CB120 [20] have been purified by the following modification of the above procedure. Infected cells (100 g) were carried through Step III using column scaled down appropriately, and the activity was assayed at 30°C. A phosphocellulose step was substituted for the DNA-cellulose chromatography since the latter was less successful with the *ts* L88 enzyme. The pooled hydroxyapatite fractions were dialyzed against a solution containing 50 mM Tris-HCl, pH 7.5, 10 mM KCl, 0.1 mM EDTA, 10 mM mercaptoethanol, and 25% glycerol (buffer E) and applied to a column of phosphocellulose (0.9 x 15 cm) that had been equilibrated with the same buffer. A linear gradient formed from 50 ml buffer E and 50 ml buffer E containing 0.4 M KCl was used to elute the enzyme, which in each case eluted at 0.15 to 0.2 M KCl. Each mutant enzyme eluted from the phosphocellulose column had a constant specific activity for both nuclease and polymerase activities throughout the column fractions, gave a single peak on disc gel electrophoresis in the presence of SDS and mercaptoethanol, and was free of contaminating DNA endonucleases using λ DNA as the substrate [19,20]. Pooled fractions from the phosphocellulose column were diluted and concentrated on a small phosphocellulose P-11 column (1 ml volume) as described for the wild-type enzyme.

C. B22 Nuclease

The gene 43 amber mutant T4 *am* B22 produces a fragment of the wild-type polymerase that has a molecular weight of about 80,000 and contains the 3' to 5' exonuclease but not the polymerase activity associated with the wild-type enzyme [3]. It has been purified by a procedure similar to that used for the wild-type enzyme [2]. The 100,000 *g* extract was prepared from the cell paste (429 g) as described for the wild-type enzyme. The enzyme was eluted from a DEAE-cellulose column (8.5 x 29 cm) at 0.04 M KCl using a linear gradient formed from 4 liters of buffer A and 4 liters of buffer A containing 0.15 M KCl. The pooled DEAE-cellulose fraction was applied to a hydroxyapatite column (6 x 15 cm) that had been equilibrated with 20 mM Tris-HCl (pH 7.5) and 0.1 M KCl. The column was washed with 200 ml buffer F [20 mM Tris-HCl (pH 7.5), 0.1 M KCl, 5 mM mercaptoethanol, 5.6 mM PMSF, and 25% glycerol] and the enzyme was eluted at 5% saturated ammonium sulfate using a linear gradient formed from 4 liters of buffer F and 4 liters of buffer F with 12% saturated (at 0°C) ammonium sulfate. The pooled hydroxyapatite fraction (296 ml) was diluted in 30-ml aliquots with 2 vol of buffer G [50 mM Tris-HCl (pH 7.5), 5 mM mercaptoethanol, 5.6 mM PMSF, and 25% glycerol] and immediately after dilution the fractions were loaded one after the other onto a phosphocellulose column (2.9 x 33 cm) that had been equilibrated with 50 mM Tris-HCl (pH 7.5). This procedure was adopted because the B22 nuclease is unstable in dilute solutions and does not withstand dialysis. The column was washed with 170 ml of buffer G with 0.05 M KCl and the nuclease was eluted at 0.10 to 0.19 M KCl using a linear gradient formed from 2 liters of buffer G with 50 mM KCl and 2 liters of buffer G with 0.8 M KCl. The pooled enzyme was diluted in 50 ml aliquots with buffer H [50 mM Tris-HCl (pH 7.5), 5 mM mercaptoethanol, and 25% glycerol] and loaded at a flow rate of

6 ml/min onto a phosphocellulose column (2 x 9.5 cm) equilibrated as above. The protein was eluted in 12 ml using buffer H with 0.5 M KCl. It has been stable for three years at -70°C. The enzyme had a specific activity of 105 units/mg using the spectrophotometric assay. It displayed constant specific activity throughout the first phosphocellulose chromatography, was free of DNA endonuclease, and gave a single band on disc gel electrophoresis at pH 8.9 that was associated with nuclease activity, and a single band on disc gel electrophoresis in the presence of SDS and mercaptoethanol.

V. PROPERTIES AND APPLICATIONS OF THE ENZYME

A. Wild Type T4 DNA Polymerase

The T4 DNA polymerase requires a primer with a 3' OH end that is hydrogen bonded to the template to be copied. When single-stranded DNA serves as the template primer, the exonuclease associated with the enzyme first hydrolyzes any unpaired regions at the 3' end of the chain until it reaches a region that is hydrogen bonded to another region of the same chain. This paired 3' terminus then acts as the primer for synthesis in the 5' to 3' direction [21]. Once synthesis begins, the addition of nucleotides to the 3' end of the DNA prevents further nuclease degradation of the original template primer. The polymerase will also copy single-stranded regions at the 5' terminus of duplex DNA, using the 3' terminus of the opposite strand as the primer [1]. Thus any duplex DNA can be labeled at its 3' terminus by first partially digesting from the 3' terminus with *E. coli* exonuclease III, followed by synthesis with the T4 polymerase. Similarly, the polymerase will copy unpaired 5' terminal regions in duplexes formed from synthetic oligodeoxynucleotides if the strand serving as primer has at least five to seven nucleotide units and the template is at least 12 units long [22]. This polymerase will repair gaps in duplex DNA completely so that the product can be sealed with the *E. coli* or T4 DNA ligase [23,24]. Unlike *E. coli* polymerase I, the T4 polymerase cannot begin synthesis at a single-stranded nick in duplex DNA apparently because it cannot displace the primer strand [25]. The T4 polymerase is stimulated in vitro by the T4 gene-32 DNA unwinding protein [26].

The exonuclease activity of the enzyme degrades denatured DNA from the 3' chain terminus randomly rather than processively (one chain at a time) yielding 5' mononucleotides and a short oligonucleotide (n = 2 to 3) originating from the 5' terminus [2]. Oligonucleotides of about 10 are degraded much faster than chains of over 200 residues [2]. Single-stranded DNA is hydrolyzed much faster than double-stranded DNA [1]. The enzyme can begin hydrolysis at a single-stranded nick in duplex DNA [1]. The exonuclease degrades denatured T4 DNA, which contains glucosylated hydroxymethyl dCMP residues [27].

During in vitro DNA synthesis by this enzyme there is extensive hydrolysis of newly incorporated residues [28]. This is easily measured by the DNA-dependent conversion of deoxynucleoside triphosphate to monophosphate. If all of the four required deoxynucleoside triphosphates are not present, the enzyme will alternately remove and insert any available triphosphates that appear at the 3' chain terminus [29]. This property of the enzyme has been exploited by Englund and his collaborators to sequence the site of cleavage by a *Hemophilus influenzae* restriction enzyme [29] and to determine the 3' terminal sequences of T7 [30] and λ phage DNA [31]. It also allows any DNA to be labeled near its 3' terminus [2].

B. T4 DNA Polymerase from Temperature-Sensitive Mutants

The polymerase from temperature-sensitive gene 43 mutants have been studied extensively, since in vivo some of these mutants increase (mutators) [32] and others decrease (antimutators) [33] the frequency of mutation of unlinked markers. The available evidence suggests that the fidelity of replication depends both upon the frequency with which the wrong nucleotide is incorporated by the polymerase and upon the ability of the exonuclease activity associated with the polymerase to remove errors made during synthesis. Thus, Hall and Lehman [34] found that enzyme from the mutator _ts_ L56 stably misincorporated dTMP in place of dGMP four times as often as the wild-type enzyme using poly(dC) as template. Another mutator enzyme, _ts_ L88, was found by Hershfield to misincorporate both dCTP and dGTP at six times the frequency of the wild-type enzyme with poly(dA·dT) as the template-primer [19]. This reaction was measured by the template-dependent formation of dCMP and dGMP since all of the residues misincorporated by either enzyme were removed by their respective exonuclease activities. Muzyczka, Poland, and Bessman have shown that some mutator enzymes have a lower ratio of nuclease to polymerase than the wild-type enzymes, while some antimutator enzymes have a higher ratio of these two activities than the wild type [35]. We find that enzyme from the antimutator _ts_ CB120 incorporates triphosphates at about the same rate as the wild-type enzyme using denatured T7 DNA as the template, but a much larger fraction of those nucleotides incorporated by the mutant are hydrolyzed during synthesis in vitro [20].

1. B22 Nuclease

The 3' to 5' nuclease associated with the amber peptide of am B22 differs from the wild type in its markedly decreased affinity for DNA and oligonucleotides [2]. It can be used to degrade the 3' hydroxyl termini of long DNA strands. With shorter chains (n = 200), DNA strands shorter than the initial substrate do not accumulate since they are hydrolyzed at a much faster rate than is the initial substrate [2]. The B22 protein will not catalyze the polymerization of DNA or the DNA-dependent conversion of any deoxynucleoside triphosphate into monophosphate, even when substrates and templates are present at high concentrations [2,28].

REFERENCES

1. M. Goulian, Z. Lucas, and A. Kornberg, J. Biol. Chem., 243, 627 (1968).

2. N. G. Nossal and M. S. Hershfield, J. Biol. Chem., 246, 5414 (1971).

3. N. G. Nossal, J. Biol. Chem., 244, 218 (1969).

4. B. M. Alberts and L. Frey, Nature, 227, 1313 (1970).

5. F. W. Studier, J. Mol. Biol., 41, 189 (1969).

6. A. V. Furano, Anal. Biochem., 43, 639 (1971).

7. N. G. Nossal and M. F. Singer, J. Biol. Chem., 243, 913 (1968).

8. I. R. Lehman, J. Biol. Chem., 235, 1479 (1960).

9. C. B. Klee and M. F. Singer, J. Biol. Chem., 243, 923 (1968).

10. A. E. Oleson and J. F. Koerner, J. Biol. Chem., 239, 2935 (1964).

11. M. H. Adams, Bacteriophages, Wiley (Interscience), New York, 1959.

12. J. S. Wiberg, M. L. Dirksen, R. H. Epstein, S. E. Luria, and J. M. Buchanan, Proc. Natl. Acad. Sci. U.S., 48, 293 (1962).

13. R. S. Edgar, G. H. Denhardt, and R. M. Epstein, Genetics, 49, 635 (1964).

14. J. Eigner and S. Block, J. Virol., 2, 320 (1968).

15. D. Fraser and E. A. Jerrel, J. Biol. Chem., 205, 291 (1953).

16. D. E. Fahrney and A. M. Gold, J. Am. Chem. Soc., 85, 997 (1963).

17. Ö. Levin, in Methods in Enzymology (S. P. Colowick and N. O. Kaplan, eds.), Vol. 5, Academic, New York, 1962, p. 27.

18. B. Alberts and G. Herrick, in Methods in Enzymology (L. Grossman and K. Moldave, eds.), Vol. 21, Part D, Academic, New York, 1971, p. 198.

19. M. S. Hershfield, J. Biol. Chem., 248, 1417 (1973).

20. F. D. Gillin and N. G. Nossal, unpublished experiments (1973).

21. P. T. Englund, J. Biol. Chem., 246, 5684 (1971).

22. K. Kleppe, E. Ohtsuka, R. Kleppe, I. Molineax, and H. G. Khorana, J. Mol. Biol., 56, 341 (1971).

23. N. Anraku and I. R. Lehman, J. Mol. Biol., 46, 467 (1969).

24. Y. Masamune, R. A. Fleischman, and C. C. Richardson, J. Biol. Chem., 246, 2680 (1971).

25. Y. Masamune and C. C. Richardson, J. Biol. Chem., 246, 2692 (1971).

26. J. A. Huberman, A. Kornberg, and B. M. Alberts, J. Mol. Biol., 62, 39 (1971).

27. W. M. Huang and I. R. Lehman, J. Biol. Chem., 247, 3139 (1972).

28. M. S. Hershfield and N. G. Nossal, J. Biol. Chem., 247, 3393 (1972).

29. P. T. Englund, J. Biol. Chem., 246, 3269 (1971).

30. P. T. Englund, J. Mol. Biol., 66, 209 (1972).

31. P. H. Weigel, P. T. Englund, K. Murray, and R. Old, Proc. Natl. Acad. Sci. U.S., 70, 1151 (1973).

32. J. R. Speyer, J. D. Karam, and A. B. Lenny, Cold Spring Harbor Symp. Quant. Biol., 31, 393 (1966).

33. J. W. Drake, E. F. Allen, S. A. Forsberg, R. M. Preparata, and E. O. Greening, Nature, 221, 1128 (1969).

34. Z. W. Hall and I. R. Lehman, J. Mol. Biol., 36, 321 (1968).

35. N. Muzyczka, R. Poland, and M. J. Bessman, J. Biol. Chem., 247, 7116 (1972).

Chapter 14

PURIFICATION OF THE GENES 44 AND 62 COMPLEX FROM T4-INFECTED *E. COLI*

Jack Barry*

Biology Department
Brookhaven National Laboratory
Upton, New York

*Present address: National Academy of Sciences, Assembly of Life Sciences, Washington, D.C.

I. INTRODUCTION

Escherichia coli cells infected with bacteriophage T4 amber mutants in genes 32, 41, 43, 44, 45, or 62 synthesize little or no DNA even though all four deoxyribonucleoside triphosphates are present [1,2]. For several of these genes, temperature-sensitive mutations exist for which DNA replication stops abruptly after a shift to the nonpermissive temperature [3,4]. These six gene products may, therefore, be directly involved in building the T4 DNA replication apparatus [4].

The products of T4 genes 43 and 32 are a DNA polymerase [5,6] and a DNA-unwinding protein [7], respectively; both of these proteins have been purified and characterized in detail [8-11]. The DNA-unwinding protein increases the in vitro rate of polymerization by the polymerase five- to tenfold [12]. In addition, these two proteins appear to specifically interact with each other [12].

In order to determine their roles in T4 DNA replication, the products of T4 genes 41, 44, 45, and 62 must also be identified and purified. For this purpose, an in vitro DNA synthesizing system that shows a requirement for these gene products was developed [13]. This system consists of concentrated, gently lysed, T4-infected cells, in which the endogenous DNA serves as the template. Like the system described by Smith, Schaller, and Bonhoeffer for *E. coli* [14] and by Okazaki and co-workers for T4-infected *E. coli* [15], these cell lysates support a brief period of rapid DNA synthesis when supplied with deoxyribonucleoside triphosphates. Using such a system as an assay, the products of T4 genes 62 and 44 have been purified to homogeneity.

II. MATERIALS AND METHODS EMPLOYED

A. Bacteria, Bacteriophage, and Enzymes

The host strain for all experiments was *E. coli* D110 (*Pol* A_1, *end* I^-, *thy*$^-$, *su*$^-$) obtained from Dr. C. C. Richardson [16]. The following T4 mutants were obtained from the Cal Tech collection: *am* HL618 (gene 32^-), *am* B22 (gene 43^-), *am* N81 (gene 41^-), *am* N82 (gene 44^-), *am* E10 (gene 45^-), and *am* E1140 (gene 62^-). In addition, T4 phage SP°/62-*am* N55 (gene ?, gene 42^-), whose SP62 genotype causes it to overproduce several early gene products when DNA replication is blocked [27], was generously provided by Dr. Wiberg. Purified T4 DNA polymerase was a gift from Dr. Wai Mun Wang; T4 gene 32 protein was prepared according to Alberts and Frey [10].

B. Preparation of Infected Cells

E. coli D110 was grown in log phase to a concentration of 4 x 10^8 cell/ml in M-9 minimal medium supplemented with 0.3% casein hydrolysate, 1 μg/ml of thiamine, and 0.02 mg/ml of thymidine. The cells were infected by addition of the appropriate mutant T4 bacteriophage at a multiplicity of infection (MOI) of 5, followed by incubation for 20 min at 37°C (for SP62-*am* N55, a 60 min incubation was used). The infected cells were harvested by centrifugation at 4°C, washed twice in 20% sucrose containing 0.05 M Tris·HCl (pH 7.4)-1 mM Na_3EDTA, and stored as a frozen pellet.

C. Preparation of Receptor Cell Lysates

A pellet of frozen cells was thawed and evenly suspended in four volumes of 25% sucrose-0.05 M Tris·HCl (pH 7.4). After addition of one more volume of 10 mM Na_3EDTA containing 2 mg/ml of egg white lysozyme (Worthington Biochemical), the mixture was incubated for 30-60 min in an ice bath. Five more volumes of buffer containing 0.05 M Tris·HCl (pH 7.4)-0.03 M $MgSO_4$ and 1% Brij-58 (Atlas Chemical Industries) were then added, and the incubation was continued for 15-30 min to complete lysis [15,17]. The final concentration of the lysed cells was about 4 x 10^{10}/ml. No special precautions against shearing were taken in handling receptor extracts.

D. Preparation of Donor Extracts

A lysate prepared as described above was centrifuged at 30,000 *g* for 15 min to pellet cell membranes and associated DNA. The supernatant was then centrifuged at 165,000 *g* for 35 min to pellet ribosomes. This second supernatant is called the "extract" (or fraction I) in our purification scheme.

E. Complementation Assay

Fifty microliters of mutant-infected receptor cell lysate is mixed with 50 μl of donor extract or buffer in an ice bath. A 25 μl aliquot is added to 25 μl of a mixture containing 0.05 M Tris·HCl (pH 7.4), 0.2 mM deoxyadenosine, 2 mM ATP, and 0.04 mM each of dGTP, dCTP, and TTP, and [^{3}H]dATP (10^5 cpm/pmole). After incubation at 37°C for 20 min, DNA synthesis is stopped by addition of 50 μl of 0.2 m Na_3EDTA and chilling. The entire mixture (100 μl) is then spotted on a glass-fiber filter (Whatman GF-A), batch-washed in 5% trichloroacetic acid and ethanol [18], and dried and counted by standard techniques. In order to quantitate activity, donor extracts are serially diluted into a buffer containing 0.05 M Tris·HCl (pH 7.4)-5 mM $MgSO_4$-1 mM β-mercaptoethanol-10% glycerol-100 μg/ml bovine serum albumin (Cal Biochem); each dilution is then tested by mixing duplicate aliquots with receptor cell lysate as described above.

III. IN VITRO DNA SYNTHESIS IN INFECTED CELL LYSATES

DNA synthesis in whole-cell lysates of T4-infected cells can be quantitated by measuring incorporation of added radioactively labeled deoxyribonucleotide triphosphates into acid-insoluble material. Such synthesis is of brief duration, lasting only a few minutes at 37°C. The product made is stable for at least 30 min of further incubation at 37°C; it can be degraded by added pancreatic DNase but not by pancreatic RNase. Figure 1 demonstrates that the rate of DNA synthesis measured in lysates decreases below uninfected values when cells are harvested shortly after infection, and then increases dramatically at later times of infection. This time course thus resembles that for in vivo DNA synthesis.

Cells infected for 20 min at 37°C were routinely used; for these cells the initial rate of synthesis is very rapid, corresponding to more than 10^3 molecules of [^{3}H]dATP incorporated into DNA per cell per second. Thus, reminiscent of the in vitro system described previously by Smith, Schaller, and Bonhoeffer for uninfected *E. coli* [14] a transient period of DNA synthesis is observed at close to in vivo replication rates.

T4 DNA replication in vivo exhibits a nearly absolute requirement for the products of T4 genes 32, 41, 43, 44, 45, and 62 [1,2]. In vitro DNA synthesis in our whole cell lysates is likewise reduced if cells have been infected with these bacteriophage mutants. As listed in Table 1, the DNA synthesis in these mutant lysates is reduced to anywhere from 20 to 1% of the wild-type level, depending on the particular gene product missing.

TABLE 1

Extent of In Vitro DNA Synthesis in Cell Lysates

Cell lysate[a]	CPM	% of wild type
Wild type T4[b]	8750	(100)
T4 gene 32⁻	1980	22
T4 gene 41⁻	1700	19
T4 gene 43⁻	120	1
T4 gene 44⁻	450	5
T4 gene 45⁻	810	9
T4 gene 62⁻	670	8
Uninfected cells	2200	--

[a]For the particular mutants used and details of the assay, see Section II. A filter blank of 50 cpm has been subtracted.

[b]The level of activity in this lysate may be misleadingly enhanced relative to other lysates since more T4 DNA template is present due to intracellular DNA synthesis.

IV. IN VITRO COMPLEMENTATION FOR DNA SYNTHESIS

In preliminary tests, purified 32 protein stimulated the amount of DNA synthesis in its deficient lysate 1.7-fold, while purified 43 protein (T4 DNA polymerase) stimulated DNA synthesis in its deficient lysate 1.5-fold. Neither protein stimulated DNA synthesis in the opposite lysate. Individual lysates deficient in the products of T4 genes 41, 44, 45, or 62 were then tested for their ability to be complemented by extracts containing these proteins. Extracts were prepared and used for complementation as described in Section II. Figure 2 (top) illustrates the level of DNA synthesis in a 62-deficient lysate after addition of varying amounts of an extract made from cells containing the product of gene 62. As shown, the donor extract could be diluted 27-fold and still evoke a twofold stimulation of DNA synthesis. However, an extract prepared in exactly the same manner from 62-deficient cells is seen to have no stimulatory effect. Similar complementation of a 45-deficient lysate by the gene 45 product is shown in Fig. 2 (bottom). Analogous complementations have been obtained with receptor cell lysates deficient in the products of genes 44 and 41, and with donor extracts prepared from any of the six mutant-infected cells listed in Table 1. Results of these cross complementations, expressed as DNA synthesis relative to a buffer control, are presented in Table 2. Note that in all cases addition of a homologous extract is without effect, while heterologous

TABLE 2

Complementation of T4 Mutant Lysates by Various Mutant Extracts--DNA Synthesis Relative to Buffer Control at 1.0[a]

T4 donor extract	T4 receptor cell lysate			
	Gene 62^-	Gene 45^-	Gene 41^-	Gene 44^-
Gene 32^-	2.7	1.6	2.1	1.7
Gene 41^-	2.1	2.2	1.0	2.3
Gene 43^-	2.3	1.6	1.9	1.5
Gene 44^-	1.2	1.6	1.8	1.0
Gene 45^-	2.1	1.0	1.5	1.9
Gene 62^-	1.1	2.0	1.6	1.2
Buffer control	(1.0 = 1800 cpm)	(1.0 = 500 cpm)	(1.0 = 2000 cpm)	(1.0 = 600 cpm)

[a]The incubations were performed with undiluted donor extracts as described in Section II, except that 3×10^9 rather than 10^{10} receptor cell equivalents/ml were used in the final assay mix. This lower cell concentration probably accounts for the rather poor complementations obtained (see Table 3).

extracts stimulate synthesis from 1.5- to 2.7-fold. This finding strongly indicates that the stimulations observed are due to addition of the missing gene product. An apparent exception to the general pattern in Table 2 is that very little complementation is observed when 44-deficient donor extracts are mixed with 62-deficient lysates, or when 62-deficient donor extracts are mixed with 44-deficient lysates. This initially puzzling observation is discussed in detail below.

Reproducible response to these gene products requires a high concentration of receptor cell lysate. For example, in the experiment shown in Table 3, DNA synthesis in a gene 62-deficient cell lysate was stimulated 2.4-fold by two extracts containing 62 protein. However, when this receptor cell lysate was diluted fourfold to 2.5×10^9 cell equivalents/ml, its DNA synthesis was barely stimulated by the same extracts. This result suggests that at least some of the factors necessary for DNA synthesis must be present at very high concentration if efficient complementation is to occur (see also Ref. 17).

TABLE 3

Effect of Receptor Lysate Concentration on the Extent of of Complementation Observed--DNA Synthesis in 62-Deficient Receptor Cell Lysate (cpm)

Donor extract	Undiluted receptor lysate[a]	Receptor lysate diluted fourfold
Buffer control	1100	870
Gene 62^-	1150	700
Gene 41^-	2640	1090
Gene 43^-	2740	1030

[a]10^{10} cell equivalents/ml in final assay mix.

V. PURIFICATION OF GENE 62 AND GENE 44 PROTEINS

Using the complementation response as an assay, 62 activity was originally purified from cells infected with T4 am B22 (gene 43^-) and later from cells infected with T4 SP62-am N55. The latter cells appear to produce 10 to 20 times as much 62 activity as the former, as estimated by complementation. This is illustrated in Fig. 3, where diluted aliquots of donor extracts from the SP62-am 55-infected cells and the am B22-infected cells (both of which are wild type for gene 62 and blocked in DNA replication) have been compared in their ability to stimulate 62-deficient lysates. Overproduction of the replication gene products by phage containing the SP62 mutation was expected from the earlier results of Wiberg [27].

The quantitation of gene 62 activity required the assay of several dilutions of the fractions obtained at each stage of purification, with a unit of activity defined as that giving half-maximal stimulation. The purification method developed consists of passing a Brij-lysed extract through a large DEAE-cellulose column; this binds all of the nucleic acids and most of the protein. The gene 62 activity, appearing in the column breakthrough, is further purified by adsorption to and elution from a hydroxyapatite column. Preparative isoelectric focusing then yields the final fraction IV, isoelectric at pH 8.2. As judged by electrophoresis on polyacrylamide gels containing sodium dodecyl sulfate, followed by protein staining [18], fraction IV consists almost entirely (>95%) of two proteins, having apparent molecular weights of 34,000 and 20,000 daltons, respectively. The recovery and purity at each step of this purification are listed in Table 4 for SP62-am 55-infected cells, with details of the procedure given in the Appendix.

As shown in Table 5, this fraction does not stimulate DNA synthesis in lysates deficient in the products of genes 32, 41,

TABLE 4

Purification of 62 Complementation Activity[a]

Fraction	Volume, ml	Protein, mg/ml	Activity, units/ml	Specific activity, units/mg	Activity recovered, %
I	100	12.4	68,000	5,000	(100)
II	120	1.23	68,000	55,000	120
III	11.5	1.14	240,000	210,000	40
IV	17.3	0.35	86,000	250,000	22

[a]A unit is defined as the amount of activity sufficient to give a 2.0-fold stimulation of DNA synthesis in the standard assay with 12.5 μl of 62-deficient cell lysate. The activities reported were determined by assay of serial dilutions of each fraction with the same receptor lysate on a single day; with these cells, concentrated fraction IV stimulated synthesis 3.1-fold. Gene 44 complementation activity copurified and gave identical yields at each step (see text). Protein was determined by biuret assay after trichloracetic acid precipitation of each fraction.

43, or 45. However, both 44-deficient lysates and 62-deficient lysates are complemented. Most significantly, as shown in Fig. 4, the percent stimulation of these two lysates is similar over a 250-fold concentration range of purified complex. It appears therefore that one of the two proteins in fraction IV is the gene 62 protein and that the other is the gene 44 protein.

Attempts were made to separate the two proteins in fraction IV by taking advantage of the difference in their size. However, on a 5-20% sucrose gradient, the proteins sedimented together as a homogeneous complex at 7.1 S [catalase (11.3 S) and *E. coli* alkaline phosphatase (6.2 S) were used as internal markers, and the gradient buffer contained 0.15 M potassium phosphate, pH 7.0, 5 mM $MgSO_4$, 1 mM β-mercaptoethanol, and 10% glycerol].

TABLE 5

Complementation of Different Mutant Lysates by the Fraction IV Protein Complex--DNA Synthesis In Vitro (cpm)[a]

Receptor cell lysate	Concentration of fraction IV protein in the reaction mixture		Stimulation of DNA synthesis
	0	4 μg/ml	
Gene 32^-	2000	2200	1.1
Gene 41^-	1700	1770	1.0
Gene 43^-	170	120	0.8
Gene 45^-	370	350	0.9
Gene 44^-	450	920	2.0
Gene 62^-	660	1740	2.5

[a]For assay conditions, see text.

With the same buffer on a Bio-Gel A 0.5 M column (BioRad Laboratories), the two proteins eluted together at an apparent (spherical) molecular weight of about 300,000 (in this case, the column was standardized with catalase, alkaline phosphatase, and myoglobin markers). Across the peak, the molar ratio of 34,000 dalton protein to 20,000 dalton protein was constant at about 2:1, as estimated by elution of bands stained with Coomassie Blue from SDS-polyacrylamide gels [19]. A combination of the sedimentation and gel filtration data [20] suggest that the complex is actually quite asymmetric (axial ratio of about 5:1 for a prolate ellipsoid), with a molecular weight of about 164,000.* The ratio

*Comparing two proteins, the Svedberg equation gives, assuming $\bar{v}_1 = \bar{v}_2$: $$M_1/M_2 = S_1/S_2 \cdot f_1/f_2$$

Our standard, beef-liver catalase, is a roughly spherical protein of 244,000 daltons that sediments at 11.3S [21,22]. Since frictional coefficients for spherical molecules are proportional to the cube root of their molecular weights, the gel filtration result implies that:

$$f_{44\text{-}62}/f_{catalase} = 1.07 \left[1.07 = (300{,}000/244{,}000)^{1/3}\right].$$

Thus, $M_{44\text{-}62} = 244{,}000(7.1/11.3)(1.07) = 164{,}000$. The same S value was obtained for the 44-62 protein complex in the buffer indicated in the text and in a buffer containing 0.02 M Tris·HCl, pH 8.1, 5mM $MgSO_4$, 1mM β-mercaptoethanol, and 10% glycerol.

of chains in this complex should therefore be 4:2 (i.e., an expected molecular weight of 176,000).

Experiments were performed in which the radioactive proteins in partially fractionated crude extracts have been analyzed by double-label techniques after pulse labeling and fractionation on SDS-polyacrylamide gels. Comparing different mutant infected cells (all of which have the SP62 genotype), we find that a protein with a molecular weight of about 34,000 is missing in a 44-deficient extract and that a protein with a molecular weight of about 20,000 is missing in a 62-deficient extract. It appears then that the heavier component in fraction IV is the product of gene 44 and the lighter component the product of gene 62.

Attempts to prove this directly by using double-label techniques to investigate the isolated 44-62 complex have been frustrated by the fact that this complex does not appear to form from its separate components with any reasonable efficiency in vitro. This is clear from the results presented in Table 6 where an extract prepared from a 1:1 mixture of SP62-am N82-infected cells (44-protein deficient) and SP62-am E1140-infected cells (62-protein deficient) has been tested for its ability to complement both 62- and 44-deficient receptor lysates. For comparison, results obtained with extracts made from each of these cells separately, and from SP62-am N55-infected cells (wild type for both 44 and 62 proteins) are also shown. It can be seen that very little 44-62 protein complex has been generated in vitro, since the level of complementation observed with the 1:1 mixed extract is only about .01 the level obtained with the in vivo generated complex. This could be due to instability or degradation of one of the components of the complex when the other is absent or to actual difficulty of complex formation. In any case, the fact that complementation requires a preformed 44-62 complex explains the weak complementation activity of 44-deficient extracts for 62-deficient lysates, and of 62-deficient extracts for 44-deficient lysates (as seen in Table 6, and earlier in Table 2)

TABLE 6

Attempt to Form the 44-62 Complex in Vitro from Its Separated Components[a]--DNA Synthesis in the Complementation Assay Relative to Buffer Control at 1.0

Genotype of donor extract	Extract dilution	62-deficient receptor cell lysate	44-deficient receptor cell lysate
	1/10	3.2	3.2
$44^{+}62^{+}$	1/100	2.2	2.9
	1/1000	2.0	2.7
	1/10	1.1	2.5
$44^{+}62^{-}$	1/100	1.1	1.0
	1/1000	1.2	0.8
	1/10	1.7	0.8
$44^{-}62^{+}$	1/100	1.1	0.8
	1/1000	1.0	0.9
$44^{+}62^{-}$	1/10	2.0	2.7
$44^{-}62^{+}$	1/100	1.3	1.5
Mixture	1/1000	1.0	0.9
Buffer control		(830 cpm = 1.0)	(470 cpm = 1.0)

[a]The mutant phage used contained the SP62 genotype, as described in the text. For the extract, equal amounts of $44^{-}62^{+}$ and $44^{+}62^{-}$ infected cells were mixed just prior to lysis. To allow time for complex formation, all extracts were incubated for 2 hr at 4°C before being assayed.

Even though the formation of the complex from its separated components could not be demonstrated in vitro, it was hoped that it might be possible to prove that the 20,000 dalton protein in the complex is the product of gene 62 by an appropriate double-label experiment. Thus, T4 *am* E1140 (gene 62^{-})-infected cells labeled with [^{14}C]leucine after infection were mixed with [^{3}H] leucine-labeled cells infected with a wild-type revertant of this phage, and fraction IV was prepared. As expected, this fraction was highly enriched for the ^{3}H isotope. However, it was anticipated that the 62-deficient lysate would exchange its ^{14}C-labeled 44 protein into the ^{3}H-labeled wild-type complex, in which case only the 20,000 dalton protein in the complex should be enriched with ^{3}H. The actual result obtained when the complex was analyzed by SDS-polyacrylamide gel electrophoresis is shown in Fig. 5. It can be seen that there are two major peaks of radioactive protein, one at 34,000 daltons and one at 20,000 daltons, and that both of these peaks are greatly enriched for ^{3}H. It appears, therefore, that the complex formed in vivo is sufficiently tight that it does not readily exchange components in vitro. Assignment of gene numbers to its two polypeptide chains must therefore remain tentative.

VI. DISCUSSION

The in vitro DNA synthesis described here is of brief duration, and its relation to true DNA replication is not know. In particular, we do not understand the wide variation in the background levels of (uncomplemented) DNA synthesis in different mutant lysates (Table 1). Perhaps several different partial reactions are possible in vitro; clearly, true replication forks need not be operative in any of our complemented lysates. Nevertheless, this system exhibits a requirement for proteins whose precise functions are unknown, but whose involvement in DNA replication has been established from the isolation of conditional lethal mutants. Consequently, we can use this system to provide an assay for isolation of these proteins. This Chapter describes the successful purification of gene 44 and 62 proteins. Dr. Hiroko Hama-Inaba has similarly purified the gene 45 protein to homogeneity using this complementation assay. As in the present case, double-label techniques confirm the identity of the isolated gene 45 protein and attest to the specificity of the assay [28].

The tight binding of gene 44 and gene 62 proteins found here, as well as the weaker interaction of gene 32 and gene 43 proteins observed previously [12], may be remnants of a much larger structure that constitutes the DNA replication apparatus in vivo. Perhaps, like the ribosome [23], this entire apparatus can be induced to self-assemble in vitro, given the proper conditions and suitable concentrations of the appropriate pure proteins and DNA. Whether gene products in addition to those tested in Table 1 will be necessary for such a reconstruction remains an open question. Although studies with known _E. coli_ DNA mutants have thus far failed to implicate host components in T4 DNA replication [24], a requirement for host RNA polymerase [25] and/or T4 "DNA delay" products [26] cannot be ruled out.

Determination of the precise role of the 44-62 protein complex in DNA replication may be difficult, inasmuch as it could require purification and characterization of all of the other replication proteins with which this complex interacts.

APPENDIX

Purification of 62-Complementation Activity

For the purification, 100 ml of extract, prepared as described in Section II from SP62 _am_ N55-infected cells, was pumped at 80 ml/hr onto a DEAE-cellulose column at 4°C (Whatman DE-52, 1 meq/g). The column, equilibrated and rinsed in 0.02 M Tris·HCl (pH 7.4)-5 mM $MgSO_4$-1 mM β-mercaptoethanol-1 mM Na_3EDTA-10% glycerol, had a packed volume 2.5 times the volume of extract and was 50 cm long. All of the 62 activity emerged in the breakthrough fractions, which were dialyzed into 0.02 M potassium phosphate (pH 7.4)-5 mM $MgSO_4$-1 mM β-mercaptoethanol-10% glycerol (fraction II). Fraction II was loaded at 45 ml/hr onto a 18 x 2.5 cm hydroxyapatite column (BioRad), rinsed with the above dialysis buffer containing 0.10 M potassium phosphate, and then eluted by elevation of the potassium phosphate concentration to 0.15 M. This 0.15 M eluate was placed in a dialysis bag, concentrated about threefold with solid sucrose, and then dialyzed against 5 mM Tris·HCl (pH 7.4)-0.1 mM $MgSO_4$-1 mM β-mercaptoethanol-10% glycerol. This material, called fraction III, was subjected to isoelectric focusing (300 V for 39 hr at 4°C), with 10% glycerol-5 mM β-mercaptoethanol added throughout the column. Eleven milliliters of fraction III was used with a 110 ml (pH 6-9) gradient in an LKB-8110 Electrofocusing column. The material was focused as recommended by LKB (LKB Produkter, Stockholm, Sweden) in a 1-40% sucrose gradient, with the cathode at the top, and was collected from the top. The pH 8.2 fraction is called fraction IV. The total amount of fraction IV isolatable from T4 *am* B22-infected cells (no SP62 genotype) was 10- to 20-fold less, with most of the additional purification needed being obtained in the isoelectric focusing step. When it was desirable to remove ampholines, fraction IV was readsorbed to hydroxyapatite and batch eluted.

ACKNOWLEDGMENTS

This research was carried out at Princeton University in the laboratory of Dr. Bruce Alberts whom I thank for his support and his many invaluable suggestions. I also thank Mr. Larry Moran and Mrs. Linda Frey for able assistance.

REFERENCES

1. R. H. Epstein, A. Bolle, C. M. Steinberg, E. Kellenberger, E. Boy de la Tour, R. Chevallez, R. S. Edgar, M. Susman, G. H. Denhardt, and A. Lielausis, Cold Spring Harbor Symp. Quant. Biol., 28, 375 (1963).

2. H. R. Warner and M. D. Hobbs, Virology, 33, 376 (1967).

3. S. Riva, A. Cascino, and E. P. Geiduschek, J. Mol. Biol., 54, 85 (1970).

4. B. Alberts, in Nucleic Acid-Protein Interactions and Nucleic Acid Synthesis in Viral Infection (D. W. Ribbons, J. F. Woessner, and J. Schultz, eds.), North Holland, Amsterdam, 1971.

5. A. DeWaard, A. V. Paul, and I. R. Lehman, Proc. Natl. Acad. Sci. U.S., 54, 1241 (1965).

6. H. R. Warner and J. E. Barnes, Virology, 28, 100 (1966).

7. B. Alberts, F. Amodio, M. Jenkins, E. Gutmann, and F. Ferris, Cold Spring Harbor Symp. Quant. Biol., 33, 289 (1968).

8. H. V. Aposhian and A. Kornberg, J. Biol. Chem., 237, 519 (1962).

9. M. Goulian, Z. J. Lucas, and A. Kornberg, J. Biol. Chem., 243, 627 (1968).

10. B. Alberts and L. Frey, Nature, 227, 1313 (1970).

11. H. Delius, N. J. Mantell, and B. Alberts, J. Mol. Biol., 67, 341 (1972).

12. J. Huberman, A. Kornberg, and B. Alberts, J. Mol. Biol., 62, 39 (1971).

13. J. Barry and B. Alberts, Proc. Natl. Acad. Sci. U.S., 69, 2717 (1972).

14. D. W. Smith, H. Schaller, and F. Bonhoeffer, Nature, 226, 711 (1970).

15. R. Okazaki, K. Sugimoto, T. Okazaki, Y. Imae, and A. Sugino, Nature, 228, 223 (1970).

16. R. E. Moses and C. C. Richardson, Proc. Natl. Acad. Sci. U.S., 67, 674 (1970).

17. H. Schaller, B. Otto, V. Nüsslein, J. Huf, R. Herrman, and F. Bonhoeffer, J. Mol. Biol., 63, 183 (1972).

18. A. L. Shapiro, E. Vinuela, and J. V. Maizel, Biochem. Biophys. Res. Commun., 28, 815 (1967).

19. E. W. Johns, Biochem. J., 104, 78 (1967).

20. L. M. Siegel and K. J. Monty, Biochim. Biophys. Acta, 112, 346 (1966).

21. R. C. Martin and B. N. Ames, J. Biol. Chem., 236, 1372 (1961).

22. D. J. DeRosier, Phil. Trans. Roy. Soc. London, Ser. B., 261, 209 (1971).

23. M. Nomura, Bacteriol. Rev., 34, 228 (1970).

24. J. D. Gross, Current Topics Microbiol. Immunol., 57, 39 (1972).

25. W. Wickner, D. Brutlag, R. Schekman, and A. Kornberg, Proc. Natl. Acad. Sci. U.S., 69, 965 (1972).

26. C. D. Yegian, M. Mueller, G. Selzer, V. Russo, and F. W. Stahl, Virology, 46, 900 (1971).

27. J. Wiberg, manuscript in preparation.

28. H. Hama-Inaba, L. Moran, and B. Alberts, in preparation.

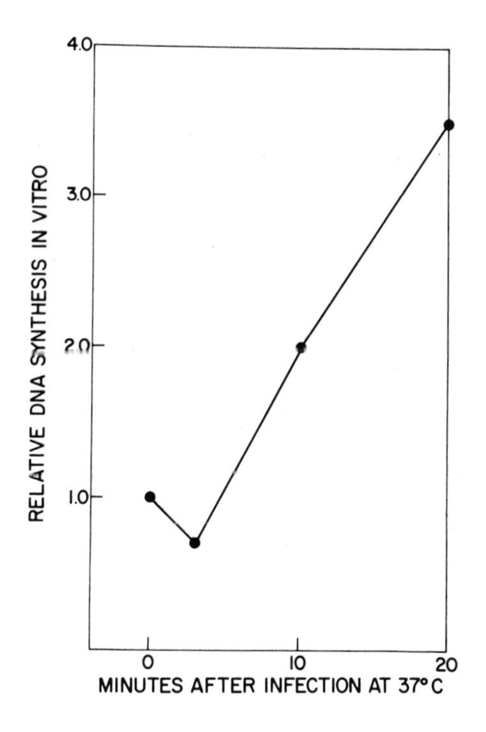

FIG. 1. DNA synthesis in vitro using cell lysates prepared from uninfected cells and cells infected with T4 am BL292 (gene 55⁻) for the indicated times at 37°C. In all cases, the lysates were incubated for 10 min at 37°C.

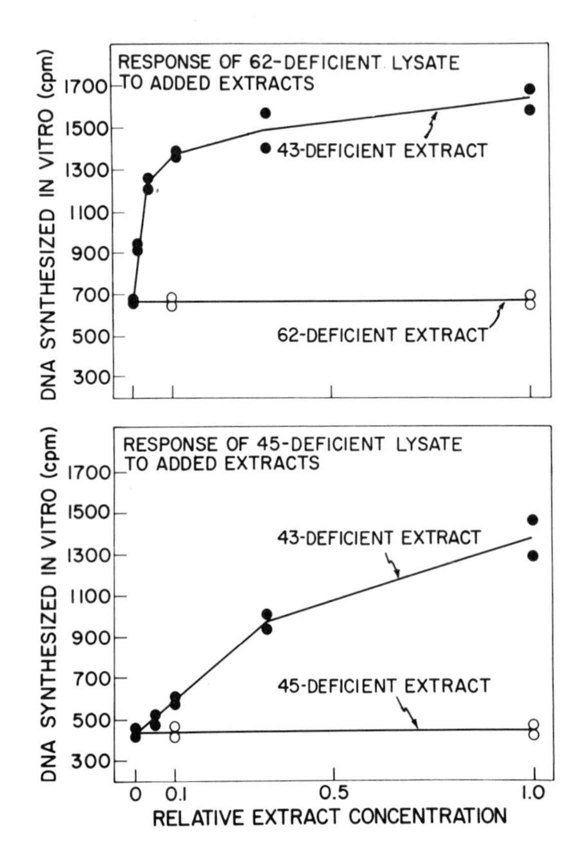

FIG. 2. Complementation of 62-deficient and 45-deficient receptor cell lysates by added donor extracts. For details of the complementation assay and phage mutants used, see Sec. II.

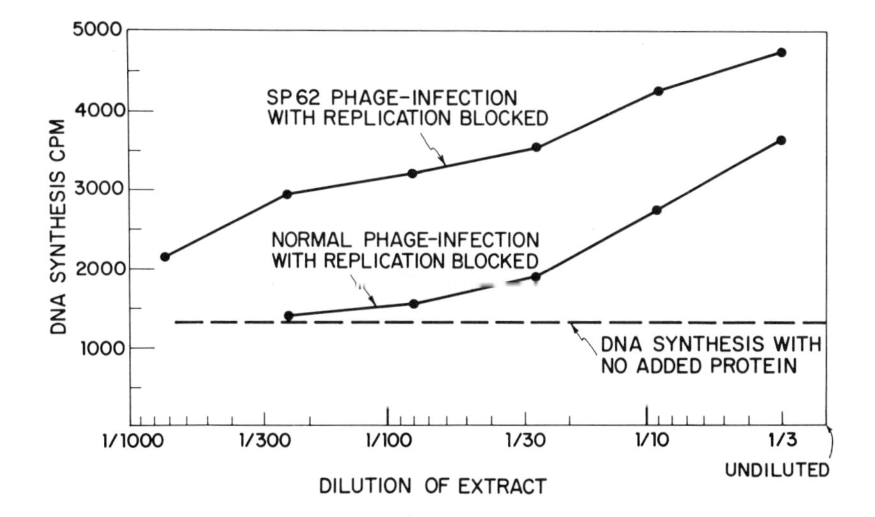

FIG. 3. Response of DNA synthesis in 62-deficient receptor cell lysates to extracts from cells infected with normal and SP62 mutant phage (both cultures infected for 40 min at 37°C). The SP62 mutation is known to cause overproduction of several T4 gene products only in the absence of DNA synthesis [27]. To stop DNA synthesis, the "normal" phage contained the am B22 mutation and the SP62 phage contained the am N55 mutation.

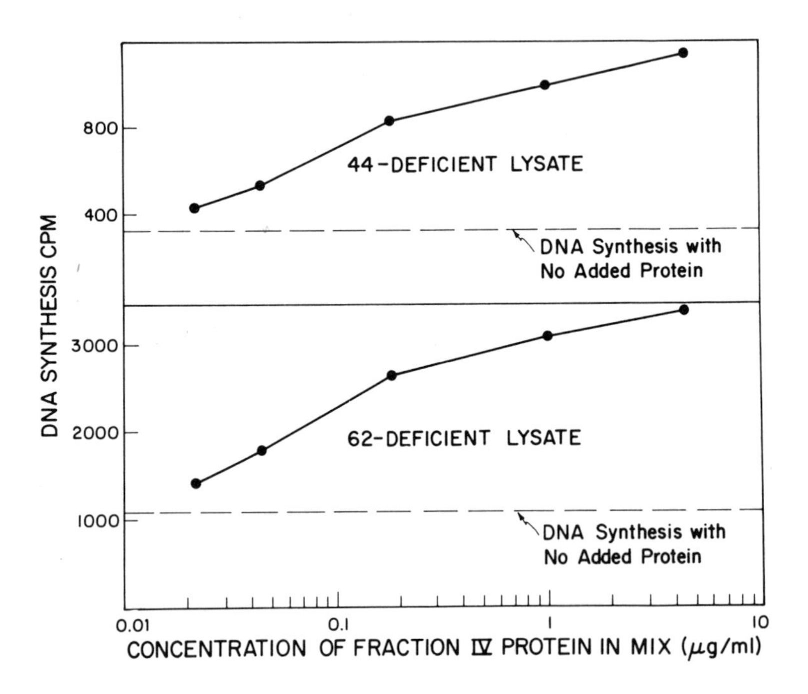

FIG. 4. Response of DNA synthesis in 44- and 62-deficient receptor lysates to dilutions of fraction IV protein. For assay conditions, see Section II.

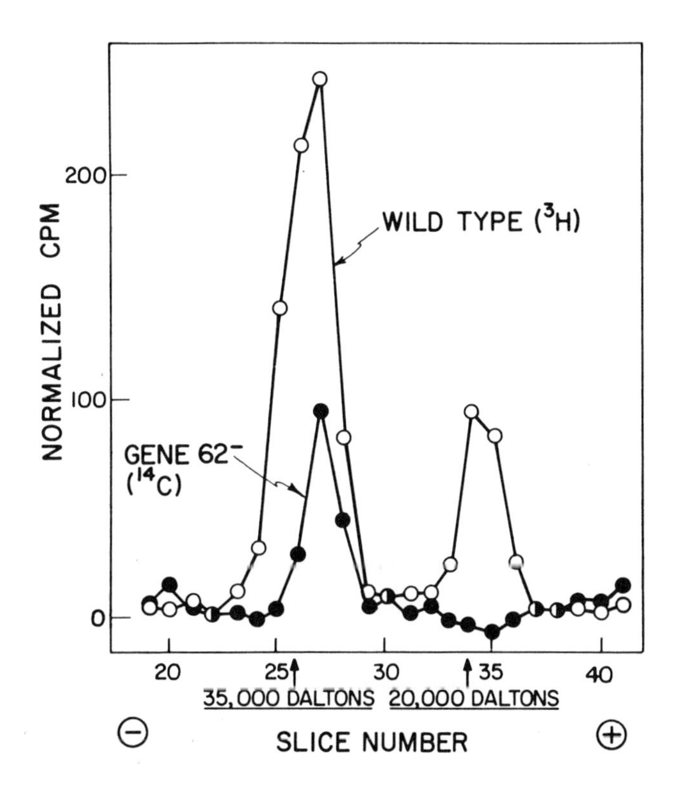

FIG. 5. Electrophoretic analysis of radioactive fraction IV proteins on a polyacrylamide gel containing sodium dodecyl sulfate. T4 am E1140 (gene 62^-)-infected cells were labeled with [^{14}C]leucine, and T4 wild-type-infected cells were labeled with [^{3}H]leucine; in both cases labeling was from 5-15 min after infection at 25°C. To ensure that the two phages differed only by a single mutation, the wild type was selected on *E. coli* B as a spontaneous revertant from the am E1140 stock. After isoelectric focusing, the proteins were precipitated with 5% trichloroacetic acid after adding lysozyme carrier and redissolved in gel sample buffer containing 1% SDS. Subsequent sample treatment, electrophoresis, and gel counting techniques were as previously described. The counts shown have been corrected for overlap and normalized to a ^{14}C:^{3}H ratio of 1.0 in the original extract.

AUTHOR INDEX

Numbers in parentheses are reference numbers and indicate that an author's work is referred to although his name is not cited in the text. Underlined numbers give the page on which the complete reference is listed.

A

B

K

L

V

W

SUBJECT INDEX

C

D

E